RIVERSIDE COMMUNITY COLLEGE
1916

P9-CKJ-883

Contents

Acknowledgments

I wish to thank many people for their valuable contributions to this report, particularly those who agreed to be interviewed or who provided written information during the researching of the book. Special thanks go to Anita Lockwood, Secretary of the New Mexico Energy, Minerals, and Natural Resources Department, for permitting part-time leave for me to research and write this book.

Once the report was written, many valuable suggestions were offered by the manuscript's reviewers, most notably Craig Matthews and Scott Kaye of Brooklyn Union Gas Company, Cecily Franklin of Consolidated Natural Gas Company, and Ashok Gupta of the Natural Resources Defense Council.

I would also like to thank the people at INFORM who contributed to the successful completion of this report: Joanna D. Underwood, President of INFORM, for her support throughout research, writing, and production, and Sibyl R. Golden, former Director of Research and Publications, for her guidance in helping create a document that is concise and clear.

Thanks also to Sharene Azimi, Editorial Assistant, for meticulous copy editing, proofreading, and photo research; to Elisa Last, Production Coordinator, for creative design and production; and to Colin Crawford, for editorial suggestions on an early draft of the manuscript.

Finally, INFORM thanks the many individuals and organizations that provided general support, and the following foundations whose grants made this study possible: The Pew Charitable Trusts, the Surdna Foundation, Inc., and The Mark and Catherine Winkler Foundation.

While the information in this book could not have been collected without the assistance of all these people, the findings and conclusions are the sole responsibility of INFORM.

Preface

We are proud to present this new INFORM report showing industry, business, and government leaders across the United States how they can respond to the nation's energy and air pollution crises by moving toward the use of natural gas in all types of vehicles – from mass transit systems to private cars.

Automotive production in the United States has increased steadily since the beginning of the 20th Century. Today, annual production is more than 55 times greater than it was eight decades ago. Of the more than 520 million cars and trucks on the road worldwide, approximately 99 percent are powered by gasoline or diesel fuel refined from crude oil. Yet oil is the most limited and rapidly depleting fuel on the planet. As supplies of domestically produced oil decline, the United States is importing about half the oil it consumes: much of it from the Middle East. As seen during the Gulf War, the economic and military costs of maintaining energy security in that region can be enormous.

Aside from the supply issue, oil-derived fuels pose serious threats to the environment and human health. Pollutants emitted by the 190 million cars and trucks in this country account for about half of all air pollution, and more than 80 percent of the air pollution in many United States urban centers. More than one-third of the nation's population lives where air quality violates federal public health standards for at least one of the key pollutants emitted by vehicles: carbon monoxide, hydrocarbons, and nitrogen oxides.

Driven primarily by energy security and environmental concerns, the United States has recently embarked on a major effort to reduce dependence on oil-derived fuels in the transportation sector. The federal Clean Air Act Amendments of 1990 and the Energy Policy Act of 1992 include comprehensive alternative fuel use initiatives. With measures ranging from setting low emissions standards for commercial fleet vehicles, to mandating federal and state government purchases of alternatively fueled vehicles, these two laws – combined with state programs – could put millions of alternatively fueled vehicles on United States roads by the end of the decade.

Given the legislative impetus for switching to alternative fuels, and the strong commitment of this new Administration, which among the competing fuels is the most viable? INFORM's 1989 study, *Drive for Clean Air*, identified a host of environmental, economic, energy security, safety, supply, and other advantages of using natural gas. *Paving the Way to Natural Gas Vehicles* further describes the relative benefits of natural gas compared to other alternative fuels, including new "reformulated" gasolines, ethanol and methanol, liquefied petroleum gas, and electricity.

A guide for rooting natural gas vehicle technology firmly in the automotive transportation world, this report – by identifying the technical, economic, governmental, and institutional areas of innovation needed for broad use of natural gas vehicles – will, we hope, lay the groundwork for a radical yet practical shift away from oil-derived fuels.

Joanna D. Underwood
President
INFORM

Chapter 1 Introduction: The End of the Gasoline Age

The one millionth automobile was built in the United States in 1912. The 500 millionth car rolled off the assembly line in 1992. Today, annual United States automotive production is more than 55 times greater than eight decades ago. Virtually all of the more than 520 million cars and trucks on the road worldwide are powered by gasoline or diesel fuel refined from crude oil.

The gasoline-powered automobile is clearly one of the spectacular technologies of the 20th century, transforming the world in unparalleled dimensions. About 48 million new vehicles are now built each year worldwide. But the end of this road is in sight. Without new motor fuels and the engine technology to use them, one more doubling of the world's automotive fleet — and, with it, one more doubling of the world's gasoline use — would strain available oil reserves, exacerbate political tensions, and dramatically increase pollution.

More than 97 percent of the energy consumed in transportation is derived from oil. More than one-third of the world's oil production is used for transportation and, in the United States, 68 percent of all oil is used in cars and trucks. Yet oil is the most limited and rapidly depleting fuel on the planet. Supplies of domestically produced oil have been declining annually for nearly a decade. Having drained much of its own resources, the United States now imports half of the oil it consumes, at an annual cost of about $55 billion. Further, on a global basis, most of the world derives its oil from a small, volatile section of the planet — the Middle East. The economic and military costs of maintaining energy security under these conditions are sizable.

Aside from the supply issue, the pollution associated with the oil now used cannot be assimilated by the environment and is endangering human life. Pollutants emitted by cars and trucks now pose one of the largest threats to the health and quality of life of the urban population in the United States, accounting for about half of all air pollution and more than 80 percent of the air pollution in many urban centers. More than one-third of the nation's population — some 85 million people — lives where air quality violates federal public health standards for at least one of the key pollutants emitted by vehicles: carbon monoxide, hydrocarbons, and nitrogen oxide.

Environmental problems associated with gasoline use also result from marine spills during the delivery of oil, from pollution emitted during the refining of oil, from groundwater contamination near leaking storage tanks at refueling facilities, and from emissions of carbon dioxide, a major contributor to global climate change.

Within the next generation, there is simply no other choice but for alternative transportation fuels — that is, fuels that are not derived from oil — to become the norm for the world's passenger vehicles.

The Move to Alternative Transportation Fuels

Driven primarily by environmental and energy security concerns, the United States has recently embarked on its first serious attempt to replace oil-derived fuels in the transportation sector. In 1989, INFORM's study, *Drive for Clean Air*, evaluated a variety of transportation fuels and presented a compelling case that natural gas is clearly superior to gasoline, methanol, and other leading alternative fuel competitors when a multitude of environmental, cost, energy security, and safety criteria are considered. Since then, the focus of the nation's alternative fuels agenda has shifted from examining the merits of alternative fuel use to actually beginning a transition from oil-derived fuels.

Spearheading this undertaking at the national level are the federal Clean Air Act Amendments of 1990 and the Energy Policy Act enacted by Congress in October 1992. Both laws include comprehensive and meaningful alternative fuel use initiatives. Together, these laws could, at a conservative estimate, put one million alternatively fueled vehicles on United States roads by the end of the decade — a total that could reach four or five million if the mandated fleet conversions are part of more broadly based efforts.

The Clean Air Act Amendments of 1990 include a variety of initiatives designed to promote alternative fuel use by vehicles. One of the most significant is aimed at government and private commercial vehicle fleets in 22 highly polluted cities, home to 31 percent of the population.[1] Beginning in 1998, 30 percent of the new automobiles purchased for centrally fueled fleets of more than 10 vehicles must meet exceptionally low emissions standards that are more stringent than those required for other communities or for other vehicles in these 22 cities. Gasoline-powered vehicles may not be able to meet these more stringent standards, but vehicles running on alternative fuels, such as natural gas, almost certainly can. By 1999, 50 percent of new fleet vehicles purchased must meet these low emissions standards, and by 2000, 70 percent must do so. Fleets of heavy-duty trucks (weighing between 8500 and 26,000 pounds) must also begin phasing in alternative fuel use in 1998, at a rate of 50 percent of new purchases.

In a separate provision, the Clean Air Act Amendments require automobile companies to manufacture for sale at least 150,000 alternative fuel vehicles in California annually, beginning in model year 1996, as part of a pilot program. The number expands to 300,000 in model year 1999. A June 1991 study by the Gas Research Institute estimates that, just to comply with the alternative transportation fuel requirements of the Clean Air Act, 602,500 natural gas vehicles could be operating by the year 2000 in the 22 United States cities targeted by the act and the State of California.

Alternative transportation fuels received a second dramatic boost with the passage of the national Energy Policy Act of 1992 on October 24, 1992. This 450-page law broadens the fleet conversion program set in the Clean Air Act. The Energy Policy Act requires the federal government to purchase at least 5000 alternatively fueled passenger vehicles in 1993; the number increases to 10,000 in 1995. In 1996, 25 percent of new vehicle purchases must be alternatively fueled, rising to 33 percent in 1997, 50 percent in 1998, and 75 percent in 1999 and thereafter. The total federal fleet of passenger vehicles is about 375,000, and annual purchases of new vehicles exceed 100,000.

The new law also establishes alternative fuel use mandates for state government vehicles operating in the more than 150 United States cities with a population greater then 250,000 people. Ten percent of new state government vehicles operating in these areas must be alternatively fueled by 1996; the percentage rises to 75 percent by the year 2000. About 2.3 million state government vehicles are affected by this program.

A third fleet mandate affecting private sector and municipal government vehicles will take effect in 1999 if the US Department of Energy determines that the program is necessary for the nation to achieve a goal of replacing 30 percent of imported motor fuels by the year 2010. More than eight million vehicles in private and municipal government fleets could be affected.

Another key provision of the Energy Policy Act provides federal tax deductions, beginning June 30, 1993, to pay the incremental cost of converting conventional vehicles to run on alternative fuels: $2000 for passenger vehicles and up to $50,000 for large trucks. In addition, the act establishes a $25 million low-interest loan fund to provide another mechanism to help vehicle owners finance conversions to alternative fuels.

State governments have also acted, on their own and often more vigorously than the federal government, to promote use of alternative transportation fuels. Since 1987, 30 states and Washington, DC have established alternative fuels programs ranging from task force studies to expansive fuel-switching mandates.[2] Some examples of state government initiatives include the following:

- California has established an increasingly stringent set of automotive tailpipe emission standards which will necessitate increasing use of alternatively fueled vehicles over the next decade (see Profile 20). Moreover, California offers a $1000 tax credit to defray the cost of converting a conventional vehicle to run on alternative fuels.
- Texas has directed qualifying state and local government car and school bus fleets to convert to alternative transportation fuels beginning in September 1994 (see Profile 15). By 1998, 90 percent of new vehicle purchases for such fleets must be able to operate on non-petroleum fuels. New Mexico, Arizona, Colorado, and several other states also have alternative fuel use mandates in place.
- Oklahoma has established a $1.5 million revolving loan fund to finance the construction of refueling stations and

conversions of vehicles to alternative fuels (see Profile 17).
Utah also runs a revolving loan program.

These government actions have prompted the genesis within the private sector of new industries that provide alternative fuel vehicles, refueling stations, and the fuels themselves. Large energy companies, the major automotive manufacturers, commercial vehicle fleet operators, and hundreds of small equipment suppliers throughout the country are entering the alternative transportation fuel business. In a few short years, this business has grown from a few million dollars per year in total sales to an annual sales base of tens of millions of dollars. Alternative transportation fuels will probably be a multibillion dollar industry by the year 2000.

Conservative estimates by the American Gas Association place the number of natural gas vehicles on the road by the year 2000 at one million, with an intermediate estimate of 50,000 vehicles by 1995. By 2005, the Association's analyses project more than four million natural gas vehicles. T. Boone Pickens, the outspoken chairman of the Natural Gas Vehicle Coalition, believes that the figure could be as high as 20 million. By any account, natural gas vehicles will be big business a few short years from now.

Natural Gas — the Best Choice

Many fuels are competing for a stake in the new alternative transportation fuels market, and virtually all government programs promoting alternative transportation fuels are fuel-neutral. The definitions of alternative fuels in these programs are broad enough to include a variety of liquid and gaseous fuels as well as electricity.

- New "reformulated" gasoline brands may reduce automotive emissions sufficiently to comply with some of the requirements of state and federal alternative fuel use provisions.
- Ethanol, distilled primarily from corn, is already mixed as a blending agent in about 8 percent of gasoline sales, somewhat reducing oil use and lessening carbon monoxide emissions.

- Ethanol and another alcohol, methanol, can be substituted for gasoline in pure form or in a mixture of 85 percent alcohol and 15 percent gasoline, with reduced tailpipe emissions. In the late 1980s, methanol was the candidate favored by federal officials and the California Energy Commission as the alternative fuel of choice in future transportation applications in the United States. Since 1990, however, concern over unexpected environmental problems and the high cost of methanol have dampened enthusiasm for methanol in many circles, and there has been an explosive surge of activity in the development and commercialization of natural gas vehicles.
- Liquefied petroleum gas (LPG), which is primarily propane, now powers about 350,000 mostly rural vehicles in the United States, making it currently the most used substitute for gasoline. There are no independent sources of LPG; it is produced only as a by-product of oil or natural gas production. Thus its supplies are limited, preventing its consideration as a widespread alternative to gasoline.
- Electricity, when used to power vehicles, leads to virtually no tailpipe emissions, although pollution can be generated during its production. However, the battery technology needed to power electric vehicles is not sufficiently developed to meet performance criteria for automobile use: the necessary batteries currently weigh about 1 ton and allow a traveling range of less than 100 miles before recharging is required.
- Solar-generated hydrogen, like electricity, can lead to virtually no tailpipe emissions when used to power vehicles. In the long term, a transition to hydrogen fuel produced from such a renewable energy source is likely for transportation applications, but the technology is not yet sufficiently developed, nor are the economics favorable enough, to permit this in the short term. (Profile 9 discusses hydrogen fuel in more detail.)
- Natural gas, however, is the most viable alternative trans-

> portation fuel, for the short term, for a host of environmental, economic, energy security, safety, supply, and other reasons. Described below, these advantages were first identified by INFORM in its 1989 study, *Drive for Clean Air: Natural Gas and Methanol Vehicles,* where a more detailed discussion of them may be found.

Although methanol and the other alternative fuels continue to draw attention, natural gas is now the front-runner in the alternative transportation fuels field, at least for the 1990s. It is the fuel of choice for many reasons: its environmental performance, lower cost, availability, safety, and proven track record; the state of technology for its use; and its ability to provide a transition to a hydrogen fuel economy.

A Clean Burning Fuel

Natural gas vehicles release up to 95 percent less air pollution from the tailpipe than gasoline vehicles. Emissions of hydrocarbons are cut by 80 percent, and nitrogen oxides are reduced by 30 percent. Together, these two pollutants interact to form ground-level ozone, or smog. Emissions of carbon monoxide, a major health problem, are slashed by 95 percent. Emissions of gases that contribute to global climate change, such as carbon dioxide, are reduced by about 15 percent compared to gasoline use.

Moreover, fuel evaporation from poorly sealed gasoline tanks and escape of gasoline fumes during refueling are eliminated with natural gas. Such evaporative emissions are the source of about half the hydrocarbon pollution from gasoline vehicles.

The production of natural gas is also less polluting than producing and refining oil or generating electricity at coal-burning power plants. Furthermore, there are no oil spills, releases of toxic pollutants from refineries, or leaks from underground storage tanks to groundwater associated with natural gas use.

All currently available alternative fuels achieve some pollution reductions, and technological advancements promise further improvements. Nevertheless, natural gas offers the best option in the near term for cost-effectively reducing automotive air pollution.

Attractive Economics

Natural gas is the only alternative fuel which, on average, is sold year-round at a substantially cheaper price than gasoline because its production costs are lower: unlike oil, natural gas comes out of the earth under pressure, and it requires no refining and very little processing. The wholesale price of natural gas is about half the wholesale price of gasoline. At the fuel pump, compressed natural gas generally sells for between half and two-thirds the price of gasoline, or between $0.50 and $0.80 per equivalent gallon (that is, the amount of natural gas containing the same amount of energy as a gallon of gasoline). Fuel savings for a heavily used car running on natural gas can exceed $500 per year, compared to gasoline; a large truck or bus can have its annual fuel costs reduced by $1500.

On the other hand, methanol, ethanol, and electricity are all at least as expensive as gasoline on an equivalent gallon basis, and can cost up to twice as much. Reformulated gasolines are expected to add between $0.10 and $0.15 to the price of a gallon of gasoline. Liquefied petroleum gas generally sells below the price of gasoline, but seasonable demand surges in winter months periodically erase this discount. Hydrogen currently costs at least three times as much as gasoline.

The savings for purchasing natural gas more than offset the additional cost of equipping a vehicle to burn this fuel. The conversion cost for vehicles capable of burning either natural gas or gasoline is about $2500-$3000 per vehicle. Fuel savings can thus recoup the capital cost of converting to natural gas in three to five years for a vehicle driven more than 20,000 miles per year. In addition, many of the parts used in natural gas vehicles, such as fuel storage cylinders, can be reused after their initial use, thereby lowering the capital cost of future conversions to natural gas.

Furthermore, over the next few years, vehicles designed and produced to burn only natural gas (dedicated natural gas vehicles) will increasingly enter the market. They are likely to cost about $1000-$1500 more than comparable gasoline-fueled vehicles. A greater portion of the fuel savings will then accrue to natural gas vehicle users.

Energy Security: Freedom from Oil

As a chemical distinct from oil, natural gas is a true alternative fuel, offering freedom from continued dependence on oil imports. Moreover, the United States currently is virtually self-sufficient in natural gas supplies. Domestic production supplies over 93 percent of the nation's demand for natural gas, and most of the remainder is supplied by Canada.

Without exploring for additional reserves, opening new wells, or constructing additional pipelines, the United States has sufficient capacity to easily fuel 20 percent of all its vehicles with natural gas — without diverting any natural gas from other current uses. And, with new markets emerging for natural gas use, including transportation, there is growing interest in developing new supplies. Further, while less drilling for natural gas has taken place in Canada and Mexico, both of these countries have significant natural gas reserves which, combined, at least equal United States reserves. Thus, natural gas offers significant security advantages compared to oil-derived fuels since the United States would not need to import natural gas from politically unstable areas or transport it across oceans.

Methanol use, on the other hand, raises energy security concerns. Although derived from natural gas, most methanol would be imported from the Middle East or elsewhere because the world's lowest cost methanol producers are overseas. All other alternative fuels mean continued oil dependence to a great extent. Reformulated gasoline, for instance, offers no energy security advantages because it is refined from oil. In fact, because of the additional refining that may be required to produce reformulated gasolines, the total oil requirement to supply reformulated gasoline may even be higher than that for conventional gasoline. Ethanol fuel blends also depend on oil for their existence since they are generally sold as a mixture of 10 percent ethanol and 90 percent gasoline (called gasohol). Although ethanol blends can extend the supply of oil-derived fuels, they do not replace the need for oil. Similarly, because about 40 percent of the world's liquefied petroleum gas is produced as a by-product of oil refining, it too is not a true alternative to oil (the other 60 percent is produced as a by-product of natural gas production, but this is an extremely inefficient production method since only three percent of the total fuel produced is liquefied petroleum gas).

Plentiful Supply

World supplies of natural gas exceed proven oil reserves by 22 percent, and there is currently greater success in finding new natural gas deposits than in discovering new oil fields. United States natural gas production has increased annually for the past eight years and now stands at its highest point since the OPEC (Organization of Petroleum Exporting States) oil embargo of 1973: about 22 trillion cubic feet per year, roughly equal to all gasoline and diesel fuel consumed in the United States. United States oil production, on the other hand, peaked in 1970 at 11.30 million barrels per day and has declined 24 percent since then, including nine straight years of production drops through 1992.

Several other alternative fuels are limited by supply constraints. For example, if all the liquefied petroleum gas produced annually in the United States were diverted from existing markets, primarily home heating, and instead burned in automobiles, it would replace less than three percent of the demand for gasoline. If every acre now producing corn in the United States were dedicated to ethanol production, the total fuel produced would supply less than 20 percent of the annual gasoline demand. Methanol is commonly produced from natural gas, with about 25 to 40 percent of the energy in the natural gas lost in the process. Thus, direct use of natural gas is better than methanol as a replacement for gasoline, from an energy efficiency standpoint.

Safety

The wide reputation of natural gas as an explosive fuel has created one of the largest barriers to its public acceptance. However, the overall safety record of natural gas, including production, delivery, and end use, is superior to that of any other automotive fuel, and is especially favorable compared to gasoline — probably the most dangerous fuel of all. Each year, more than 1700 people in the United States die from burns inflicted by gasoline ignited during accidents. By all measures of vehicle safety, natural gas vehicles are safer to operate than gasoline-powered vehicles.

The most recent national survey, published in March 1992 by the American Gas Association, found that the injury rate in accidents involving natural gas vehicles, measured in terms of vehicle miles traveled, was 37 percent lower than the rate for gasoline vehicles. Furthermore, based on records of 278.3 million miles traveled, no deaths involving natural gas vehicle use occurred, compared to a death rate for gasoline vehicles of 2.2 deaths per 100 million miles traveled. Profile 21 provides further details about a study of the comparative risks posed by accidents involving natural gas and gasoline-powered vehicles.

Despite the fact that natural gas is generally stored on board vehicles at a pressure of between 3000 and 3600 pounds per square inch, the safety record of natural gas storage cylinders far surpasses that of gasoline fuel tanks. There has never been a death from an exploding fuel storage cylinder during the 25 billion miles natural gas vehicles have been driven worldwide.

The clear reason why natural gas vehicles have resulted in fewer passenger injuries (and no deaths) is that fires are less likely to occur in crashes involving natural gas vehicles than in those involving gasoline-powered vehicles. Should a fuel leak occur, natural gas disperses quickly into the atmosphere. Liquid fuels or propane, by contrast, form highly flammable puddles. Furthermore, the ignition temperature of natural gas is much higher than that of gasoline (1300°F versus 800°F), and natural gas burns only when its concentration in air is within a narrow range of 5 to 15 percent.

The experience during an accident in Warminster Township, Pennsylvania in early 1992 attests to this. A 1991 model year police car converted to run on natural gas was rear-ended by a dump truck driven by a drunk driver. The police car was then jolted again in a chain collision as the dump truck itself was rear-ended by another truck (also driven by a drunk driver). The police car was destroyed and the police officer was severely injured in the accident. The gasoline tank ruptured, spilling fuel onto the ground (fortunately, without igniting), but no damage to the natural gas storage system occurred, including fuel leakage or fire. The Warminster police department has since converted another ten vehicles to natural gas.

Natural gas is also safer to transport to refueling stations than gasoline. There are more than 1.1 million miles of natural gas pipelines in the United States, serving nearly three-quarters of the population. Industrial safety statistics show that the accident rate for delivery of natural gas by these pipelines to end users is one thousand times lower per unit of energy (millions of BTUs) than the accident rate for gasoline delivery by tanker truck. (These statistics do not evaluate the relative safety of end uses in gas stations, homes, or vehicles.)

Natural gas and gasoline are both much safer than methanol, which poses a variety of health and safety risks when used as a transportation fuel. When ingested or absorbed through the skin, methanol is at least twice as toxic as gasoline. Because its vapor is slightly heavier than air, pockets of released methanol vapors can create explosive conditions, pose fire hazards, and cause other health threats. Further, since methanol is water-soluble, when spilled it can contaminate surface or groundwater more rapidly than water-insoluble gasoline or diesel fuel.

Established Technology

Using natural gas as a transportation fuel poses no major unsolved technical problems. Natural gas vehicle technology is at least as sophisticated as the automotive technologies available for using ethanol- and methanol-based fuels or liquefied petroleum gas, and is far more advanced than electric or hydrogen vehicle technology.

Conventional gasoline-powered vehicles can easily be converted to burn natural gas as well as gasoline: the engine remains largely the same, with most of the changes involving only the fuel storage and delivery systems. These "bi-fuel" vehicles can generally be driven 150-200 miles on natural gas before it is necessary to refuel or switch to gasoline. Diesel engines have also been successfully modified to burn natural gas.

In 1992, the major domestic automotive manufacturers began producing vehicles designed to burn only natural gas. The engines of these "dedicated" natural gas vehicles, while essentially the same as gasoline engines, have been engineered to take full advantage of the more efficient burning characteristics of natural gas; their performance generally meets, and by some measures exceeds, that of gasoline-powered vehicles. Nevertheless, more research is needed to fully optimize natural gas engine performance.

One of the greatest technological challenges involves reducing the weight and space requirements for storage of natural gas under high pressure. As noted in the above section on safety, current storage cylinders, which are constructed of steel or aluminum and are installed in the trunk or bracketed to the chassis, meet very rigorous safety requirements. However, they add 200-400 pounds to the weight of a vehicle and are about four times bulkier than gasoline storage tanks.

Refueling technologies are also established. Natural gas vehicles can be refueled overnight using small compressors that gradually pump natural gas into the storage cylinders. This is called "slow-fill" refueling. Alternatively, in "fast-fill" stations, larger compressors can be used to refill a natural gas vehicle's fuel storage cylinders in three to five minutes. As of late 1992, there were more than 500 natural gas refueling stations operating in the United States, and new ones were opening at the rate of two or three per week.

A Proven Track Record

Natural gas was used as an automotive fuel in the nineteenth century, and about 700,000 natural gas vehicles now operate worldwide. Six countries have more than 30,000 natural gas vehicles each: Italy, the former Soviet Union, New Zealand, Argentina, the United States, and Canada.

The first major natural gas vehicle commercialization effort occurred more than 50 years ago in Italy. In the late 1930s, more than 100,000 cars were operating on natural gas in Italy; today there are still more natural gas vehicles in Italy (250,000) than in any other country.

In the 1980s, New Zealand and Canada undertook major alternative transportation fuel programs focused on the increased use of natural gas. In late 1992, there were about 60,000 natural gas vehicles on New Zealand's roads, and about 35,000 in Canada. Natural gas vehicle use in the United States also began in earnest in the 1980s; there are now approximately 30,000 of them in this country.

Argentina is one of the newest and currently the fastest growing natural gas vehicle market in the world. Between 1990 and 1992, about 75,000 Argentinean vehicles were converted to natural gas use, mainly in the Buenos Aires region. There are reportedly about 200,000 natural gas vehicles operating in the countries of the former Soviet Union.

More vehicles on United States roads today are powered by natural gas than by any other alternative fuel except liquefied petroleum gas. Experience with vehicles powered by pure ethanol and methanol is limited to a few thousand cars and trucks. There are even fewer full-size electric and hydrogen vehicles.

Transition to a Hydrogen Fuel Economy

Perhaps the most valuable contribution natural gas can make to this country over the long term is to provide a clean-burning transition to an energy economy ultimately based on totally non-polluting hydrogen. Hydrogen produced from renewable resources could eventually provide pollution-free energy, but this technology is not yet viable. As a gaseous fuel, natural gas use in automobiles involves very similar technology to hydrogen use. Moreover, fuel distribution and dispensing systems would also be similar. The transition from natural gas to hydrogen is described in more detail in Profile 9.

Notes

1. The cities include all areas designated by the US Environmental Protection Agency as extreme, severe, or serious violators of the federal public health standard for ozone air pollution, plus one city (Denver) which was added to the list because of its high levels of carbon monoxide air pollution. The 22 cities are: Extreme ozone: Los Angeles; Severe ozone: Baltimore, Chicago, Houston, Milwaukee, New York City, Philadelphia, San Diego; Serious ozone: Atlanta, Bakersfield (CA), Baton Rouge, Beaumont (TX), Boston, El Paso, Fresno, Hartford, Huntington (WV), Providence, Sacramento, Springfield (MA), Washington, DC; Carbon monoxide: Denver.

2. States with alternative fuels programs as of late 1992: Arizona, Arkansas, California, Colorado, Connecticut, Florida, Georgia, Hawaii, Iowa, Kentucky, Louisiana, Maryland, Massachusetts, Minnesota, Missouri, Nevada, New Mexico, New York, North Carolina, Oklahoma, Oregon, Pennsylvania, South Dakota, Tennessee, Texas, Utah, Virginia, Washington, West Virginia, and Wisconsin.

Chapter 2 A Road Map for Change

Like any new technology, natural gas vehicles face significant barriers to widespread acceptance and use. To take just one example, the awesome size of the existing automotive and oil industries, with their allegiance to gasoline, presents an especially difficult challenge. One out of six jobs in the United States relies on the automotive industry. It is a $375 billion a year industry, with huge vested interests and a $200 million a year advertising budget. In the time it takes to snap your fingers twice, about one second, automobiles and trucks in the United States burn approximately 4000 gallons of fuel. In the time it takes to read this sentence, about 10 seconds, United States motorists have driven more than 700,000 miles. However, despite this industry's vast scope, it is showing increasing interest — spurred by changes in federal and state laws — in designing and producing alternative fuel vehicles. (This interest is detailed in Profile 2.)

Paving the Way to Natural Gas Vehicles: 25 Steps Needed

The obstacles confronting natural gas vehicles are both finite and surmountable. This book identifies the principal barriers and shows how these can best be overcome. It discusses 25 specific steps that need to be taken to help root fledgling natural gas vehicle efforts firmly in the automotive transportation world. The steps are divided into four groups: research, development, and demonstration; commercialization and refueling infrastructure development; government incentives; and removal of institutional and regulatory barriers. Each of the 25 actions is instrumental to the proliferation of natural gas vehicle technology.

For each of the steps, *Paving the Way to Natural Gas Vehicles* discusses what the action is, explains why it is important to the proliferation of natural gas vehicles, and examines some of the most promising efforts currently underway to achieve the desired goal. Each of these profiles highlights one or more organizations in its title and devotes the bulk of the profile to a discussion of the work these groups are doing. However, in many cases, smaller, but still significant, related efforts are also taking place, and these too are covered in the discussion.

The natural gas vehicle field is in constant flux. New developments are taking place while you are reading this sentence. As of late 1992, when this book went into production, every effort had been made to ensure that the profiles were as up-to-date as possible. It is inevitable, however, that changes have taken place since then.

Summary of the Profiles

The 25 profiles in this book describe the 25 steps identified by INFORM as crucial for paving the way to natural gas vehicles. They represent a road map for changing the global automotive status quo.

As noted above, these profiles are grouped into four broad categories of initiatives: research, development, and demonstration; commercialization and development of a refueling infrastructure; government incentives; and removal of institutional barriers. The individual profiles in each category are summarized below.

Research, Development, and Demonstration

Without technology optimized for natural gas use and capable of performing as reliably as conventional gasoline vehicle technology, natural gas vehicles will never be more than a curiosity in the marketplace. Hence, the first nine profiles involve research, development, and demonstration of new automotive technology that utilizes natural gas.

Profile 1. Basic Natural Gas Vehicle Engine Development: The Gas Research Institute The internal combustion engines used in the 190 million vehicles on United States roads come in a wide range of sizes and designs, from the four-cylinder, spark-ignition gasoline engines used in compact cars to the heavy-duty, compression-ignition diesel engines used in transit buses and trucks. Modifying and perfecting each engine type for efficient natural gas use requires extensive engineering research and development.

Profile 1 describes some of the dozens of projects — involving light-, medium-, and heavy-duty engines — funded by the Gas Research Institute, the research arm of the natural gas industry and the largest financial supporter of engine development research. Through the contracts awarded by the Gas Research Institute, several dozen companies have initiated or expanded natural gas vehicle projects. The positive results of the funded projects have prodded other companies to enter the natural gas vehicle field.

Profile 2. Design and Assembly Line Production of Natural Gas Vehicles: The Big Three Auto Producers The Big Three United States auto producers (General Motors, Ford, and Chrysler) produce about one-third of the world's new cars (including passenger sedans and light-duty pick-ups, wagons, and vans), and about 92 percent of those built in the United States. The reluctance of the Big Three to embrace alternative transportation fuels has historically been one of the largest obstacles to alternative fuel use, and a serious commitment on their part to assembly line production of natural gas vehicles is essential if natural gas vehicles are to become a major component of the United States automotive market. In the past few years, changes in federal and state laws have spurred the Big Three to begin taking an active interest in the design and production of alternative fuel vehicles. Profile 2 discusses their natural gas vehicle efforts.

Profile 3. Light-Duty Fleet Demonstrations: Gas Utilities and Disneyland Commercial fleet vehicles are ideally suited for demonstrating the economic and other advantages of natural gas vehicles. They are in constant use, so the owners quickly reap the benefits of the lower price of natural gas (compared to gasoline); a majority return each day to a central location, permitting overnight refueling; and they typically travel less than 200 miles each day, the distance after which, with current technology, natural gas vehicles require refueling. Further, since most of the 11 million cars and light-duty trucks in fleets of 10 or more are in urban areas, the use of natural gas vehicles in such fleets can have a significant impact on reducing air pollution in heavily polluted urban areas.

Profile 3 describes a fleet demonstration project underway at the Northern Indiana Public Service Company that is saving the company $250,000 in annual fuel costs, and a long-term demonstration program at Disneyland, which has been using natural gas vehicles to power its rides and internal transportation system since 1968.

Profile 4. Medium-Duty Commercial Fleet Demonstrations: United Parcel Service and Federal Express From an economic standpoint, trucks present an especially suitable market for natural gas vehicle development and demonstration because they have high fuel consumption rates: a typical truck uses more than 1300 gallons of fuel annually, nearly three times the 500 gallons consumed by the average car. Additionally, natural gas-powered trucks can play a vital role in ensuring energy security since they could continue to provide vital services, such as food and goods delivery, if imported oil supplies were interrupted. Medium-duty trucks (such as commercial delivery vans), weighing between 6000 and 10,000 pounds, comprise about 24 percent of the 43 million trucks on United States roads. They are beginning to receive significant attention for natural gas vehicle demonstration programs.

Profile 4 describes a United Parcel Service initiative, now in effect in six cities (Dallas, Los Angeles, New York City, Oklahoma City, Tulsa, and Washington, DC). The program has demonstrated lower maintenance and fuel costs, greater fuel efficiency, and substantial decreases in emissions of hydrocarbons, nitrogen oxides, and carbon monoxide. The profile also discusses a similar program that Federal Express is conducting in Los Angeles.

Profile 5. School Bus Demonstrations: Garland and Harbor Creek School Districts and California School bus fleets offer another promising opportunity for using natural gas vehicles: the more than 500,000 school buses represent more than 81 percent of all United States buses; their driving patterns are suited to natural gas use; their low gasoline mileage and consequent high gasoline consumption enable school districts to save money with cheaper natural gas; and their high road clearances provide ample room for safe installation of natural gas storage cylinders. Moreover, since use of any fuel in school buses is a concern because they are primarily used to transport children, the successful use of natural gas in school buses can help provide reassurance about the safety of natural gas vehicles.

Profile 5 describes demonstration projects in school districts in Texas and Pennsylvania. In the Garland, Texas, project, which now involves more than 90 school buses, the school district has more than recouped its $389,773 capital investment through fuel cost savings of more than $100,000 annually and reduced maintenance expenses of more than $4000 annually. The profile also discusses school bus demonstration programs sponsored by California and the federal government.

Profile 6. Urban Bus Demonstrations: The Federal Transit Administration Alternative Fuels Initiative From an environmental standpoint, urban buses present an ideal opportunity for natural gas conversion because they are egregious air polluters: an urban bus releases 500 times more particulate emissions than an average passenger car. The 1990 Clean Air Act Amendments mandate major reductions in these emissions beginning in 1993. However, most of the nation's 58,000 urban buses are powered by diesel engines, which present more technical obstacles to natural gas conversion than gasoline engines.

Profile 6 discusses these technical obstacles, three approaches to resolving them, and a pioneering demonstration project undertaken by Brooklyn Union Gas Company. It then describes the largest federal alternative transportation fuels demonstration program involving heavy-duty vehicles, which offers federal funds to cities for the purchase of urban buses that reduce pollution from diesel smoke. As of mid-1992, more than 200 buses were on the road as a result of this program (more than half of them fueled by natural gas), and the Federal Transit Administration had approved the purchase of an additional 300 vehicles.

Profile 7. Liquefied Natural Gas in Vehicles: Houston Metro and Roadway Express A major drawback of current natural gas vehicle technology is that, by comparison to gasoline storage systems, conventional natural gas storage cylinders are heavy and bulky because the compressed gas is stored at high pressure. These problems are significantly reduced by an alternative method of natural gas storage, the use of liquefied natural gas (LNG). Once liquefied, natural gas can be stored at lower pressures in tanks that are only half the size and weight of compressed gas tanks holding the equivalent amount of fuel.

Profile 7 describes the technical, economic, and safety challenges that must be met before the use of LNG is widespread, as well as pioneering LNG-powered vehicle projects in San Diego, Canada, and Australia. The profile then examines two ambitious United States efforts: Houston Metro's decision, following a demonstration project that revealed size and weight advantages from using LNG rather than compressed natural gas, to convert more than 300 of its transit buses to LNG use, making it the country's largest alternatively fueled transit bus fleet; and a LNG testing program conducted by Roadway Express, a company that operates 3500 long-haul moving vans and 5800 local moving trucks that annually consume 85 million gallons of diesel fuel.

Profile 8. Alternative Fuel Storage Technology: Institute of Gas Technology As noted in the description of Profile 7, the high pressures under which natural gas is typically stored necessitate storage cylinders that can weigh as much as 500 pounds (compared to just over 100 pounds for conventional gasoline storage tanks) and can occupy about four times the volume of a gasoline tank. This weight causes a net reduction of fuel efficiency in natural gas vehicles, and the cost of compressing natural gas adds about $0.20 to the price of an equivalent gallon of the fuel (that is, the amount of natural gas that produces the same energy as a gallon of gasoline).

Profile 8 describes the strategies used to date to overcome these size and weight problems: limiting the amount of natural gas storage (and hence the operating range of the vehicle) or using both natural gas and another fuel in the same vehicle. It then discusses more recent advances in storage technology and examines in detail a promising strategy that is in an early research stage and is years away from commercial application. Although it may seem counterintuitive, a cylinder that has been packed with certain types of activated carbon materials can hold a greater amount of natural gas at a lower pressure than can a cylinder packed with natural gas under high pressure but with no carbon adsorbents. The Institute of Gas Technology has been conducting the most extensive United States research program into these adsorbent natural gas storage systems.

Profile 9. Natural Gas and Hydrogen Fuel Mixtures: The Denver Hythane Project Despite the considerable benefits of natural gas as a transportation fuel, its supply is limited and nonrenewable, and its widespread use, since it is gaseous rather than liquid, would require substantial changes in the nation's infrastructure. However, these drawbacks may not pose long-term problems if a transition to natural gas vehicles helps pave the way for the use of another gaseous fuel — the virtually inexhaustible and pollution-free hydrogen fuel derived from solar energy. That is, in order to use natural gas vehicles, infrastructure and technologies may be developed that will also be necessary for the ultimate utilization of hydrogen.

While dedicated hydrogen vehicles are not yet widely available, and their technology is not yet commercially viable, Profile 9 discusses a currently available alternative to the use of pure hydrogen. A Denver project, designed to evaluate the performance of vehicles fueled by a mixture of 15 percent hydrogen and 85 percent natural gas (called hythane), is the first major United States project testing the use of such fuel mixtures in vehicles operating under routine driving conditions. It has demonstrated impressive reductions in exhaust pollution even with very small quantities of hydrogen: one-half the hydrocarbon emissions and one-third the nitrogen oxide emissions of pure natural gas.

Commercialization and Refueling Infrastructure Development

Once research and demonstration programs show that natural gas vehicles are viable from technological and economic vantage points, the next challenge is creating market demand sufficient to bring about widespread commercial natural gas vehicle production. This, in turn, requires the creation of an appropriate refueling infrastructure. There is, in fact, something of a "chicken and egg" syndrome. On the one hand, automotive manufacturers are reluctant to invest in commercial scale natural gas vehicle production due to low consumer demand, which is itself restrained by the fact that it is difficult, if not impossible, for most private individuals to have access to a natural gas vehicle refueling station. On the other hand, investments in refueling stations are held back by the lack of natural gas vehicle drivers to patronize them. The five profiles in this section describe actions that are being taken to stimulate commercial demand and establish a refueling infrastructure.

Profile 10. Developing Large Purchase Orders for Natural Gas Vehicles: Ten-Utility Natural Gas Vehicle Consortium and Federal Efforts Alternative fuel research programs typically involve only a few test vehicles, and demonstration programs commonly showcase only a limited number of vehicles. Full-scale commercialization of natural gas vehicles will only be achieved when thousands of these vehicles are available for sale in large United States automotive sales markets. A logical way to begin building a market for natural gas vehicles is by encouraging large purchase orders. Large purchase orders for natural gas vehicles during the next few years will both support the continued economic viability of manufacturing kits to convert conventional gasoline-powered vehicles and encourage automotive manufacturers to retool for assembly line production of natural gas vehicles. Such orders will promote economies of scale and lower the cost of each vehicle.

Profile 10 describes a successful effort to build commercial demand for natural gas vehicles by a consortium of ten natural gas utility companies in California, Colorado, and Texas. Together, they negotiated a purchase order with General Motors for 1000 dedicated natural gas-burning Sierra pickup trucks; GM later announced that it planned to market up to 10,000 of these trucks by the end of the 1992 model year. The profile also discusses federal efforts to generate large purchase orders through federal and state government vehicle procurement programs.

Profile 11. Assuring Reliable Aftermarket Conversion Kits: American Gas Association Laboratories In the short term, before mass production of dedicated natural gas vehicles, wider availability of natural gas vehicles depends upon an "aftermarket" that converts vehicles manufactured to burn gasoline or diesel fuel into vehicles capable of burning natural gas. While the basic engineering principles on which conversion systems are based are the same, parts and designs of conversion kits are not standardized. The number of product differences inevitably results in incompatible designs and a wide range of performance characteristics. The availability of conversion kits of varying quality presents a problem for widespread commercialization of natural gas vehicles; it is necessary to provide buyers of conversion kits with some form of quality guarantee. To fill this need, natural gas conversion kit certification programs are being established, and educational institutions are developing training programs for mechanics and purchasers.

Profile 11 describes the major nationwide certification program spearheaded by the American Gas Association Laboratories: it includes construction, performance, and manufacturing components. By mid-1992, five conversion kit manufacturers had completed the certification process, and a variety of manufacturers of component parts (e.g., regulators, valves, fuel mixers, storage cylinder brackets) used in conversion kits had had the parts certified. The profile also discusses state certification programs, as well as several specialized vocational education programs that focus on natural gas vehicle technology.

Profile 12. Establishing a Refueling Station Infrastructure: Natural Fuels Corporation and the Natural Gas Vehicle Zone Although the lack of a refueling station infrastructure has presented a significant obstacle to the widespread commercialization of natural gas vehicles, creation of such an infrastructure is now underway. Currently, there are two basic types of refueling systems: fast-fill systems, which can replenish natural gas vehicle storage cylinders in three to five minutes but are costly to build and operate, and slow-fill systems, which are much less costly but take six to twelve hours to refuel a storage tank.

Profile 12 describes the emerging refueling infrastructure. As of mid-1992, some 511 natural gas refueling stations were operating in 43 states and the District of Columbia. Minneapolis-based Minnegasco, owner of the largest utility-owned station open to the public, expects to recoup the $500,000 cost of its fast-fill station in less than two years through natural gas sales at $0.60 per equivalent gallon. The profile then focuses on one of the most ambitious private sector natural gas refueling initiatives. Undertaken by the Natural Fuels Corporation, headquartered in Denver, the project plans to establish 80 refueling stations along Colorado's Front Range by 1996. The profile also describes a plan by seven southwestern states to create a network of retail natural gas refueling stations, called the Natural Gas Vehicle Zone, along interstate highways from Louisiana to California.

Profile 13. Demonstrating Vehicle Refueling Appliances: FuelMaker Corporation Natural gas is the only alternative fuel with the potential to offer consumers freedom from the corner service station. Vehicle refueling appliances, a technology now on the market, allow individual consumers to refuel natural gas vehicles overnight, while a vehicle sits in a driveway or garage at a home or business. These refueling appliances use natural gas tapped from home or business hook-ups which, in turn, are connected to the more than 1.1 million miles of natural gas pipelines that deliver the fuel across the United States to locations where about 95 percent of the population lives and works. The creation of an individual refueling network serving even a small segment of this market could have a revolutionary impact on the country's refueling patterns.

Profile 13 describes the FuelMaker, the pioneering vehicle refueling appliance — how it works and how much it costs. Just 28 inches high and long, and weighing only 145 pounds, the FuelMaker uses a small compressor similar in size to those in large home refrigerators for slow-fill refueling. The profile also discusses the potential need for utilities to bill separately for gas used for vehicles as opposed to gas used for heating, cooking, etc., and notes the short-term option of mobile refueling.

Profile 14. Removing Restrictions against Retail Natural Gas Sales: Federal and State Governments One of the largest obstacles to commercialization of natural gas vehicles and establishment of a refueling infrastructure involves state and federal laws that prevent private sector entities from operating natural gas refueling stations. These so-called "sale-for-resale" prohibitions were originally designed to prevent price gouging by apartment house landlords. They declare anyone wishing to resell natural gas to be a public utility, and thereby subject to extensive reporting requirements, lengthy rate approval and review procedures, and observance of rigid price controls. Where sale-for-resale laws are in effect, they can stop the development of a natural gas refueling infrastructure. This obstacle, however, can be overcome at any time, and at virtually no cost, simply by convincing legislators of the merits of using natural gas as a transportation fuel and removing sale-for-resale restrictions for this use.

Profile 14 notes that eleven states have removed sale-for-resale restrictions, and explains how they have done so. It also discusses Federal Energy Regulatory Commission rules governing interstate natural gas sales which exempt transportation-related uses from federal sales regulations.

Government Incentives

The size and established infrastructure of the nation's automotive transportation system present formidable barriers to the introduction of technological innovations; indeed, the automotive and oil industries have historically resisted important technological changes (the universal installation of seatbelts is just one example). In the face of this resistance, government incentives for a transition to natural gas vehicle production are needed. Government intervention in the energy marketplace is nothing new: for example, the oil industry grew as a result of government incentives and, for the past decade, the ethanol industry has been a major beneficiary of generous government tax preferences. The many advantages of natural gas as a transportation fuel argue in favor of creating comparable incentives for natural gas use, or at least taking action to place it on an equal footing with other fuels. This group of five profiles includes both government mandates and government inducements.

Profile 15. Alternative Fuel Use Mandates: Texas and Other States
To create a large, steady market demand for alternative fuel vehicles, the US Congress and 13 state legislatures have enacted mandatory alternative transportation fuel use programs. The hope is that this demand will stimulate production of alternative fuel vehicles by the major automotive manufacturers and encourage the construction of appropriate refueling facilities. Although the programs are fuel-neutral (that is, the mandates can be filled through the use of any of a number of fuels), natural gas is the front-runner in virtually every case.

Profile 15 discusses the Texas effort, the first and still most ambitious mandatory alternative fuel use program in the country. The program requires alternative fuel use in state government fleets containing more than 15 vehicles, school bus fleets larger than 50 buses, and all metropolitan and city transit buses; if this proves insufficient to ensure that air quality meets federal public health statutes, a separate statute mandates alternative fuel use in nearly all local government and private sector fleets. As a result of the Texas program, scores of fleet conversions are underway, and dozens of natural gas refueling stations are under construction. In addition, the profile highlights key features of other state programs.

Profile 16. Grant Programs: New Mexico and the Urban Consortium Alternative fuel use mandates are undoubtedly the biggest "sticks" that a government can wield to promote a switch from reliance on oil-derived fuels. By contrast, direct monetary awards represent large "carrots" available to a government wishing to promote this switch. In fact, since 1990, state governments have greatly expanded programs aimed at encouraging the use of alternative transportation fuels; local governments have also initiated similar programs.

Profile 16 highlights the $1.8 million transportation grant program in New Mexico, one of the first states to offer significant grants for natural gas vehicle projects. As of October 1992, the program had awarded $1.8 million to fund 176 natural gas vehicle conversions. Among the projects funded are bus purchases, pilot school bus conversions, and curriculum development. The profile also discusses grant programs in New York and Pennsylvania and a major effort coordinated by the Urban Consortium Energy Task Force, under the general oversight of the National League of Cities, to promote natural gas vehicle use in cities. Funded by the US Department of Energy, these municipal projects are taking place in Houston, Pittsburgh, Denver, Albuquerque, and San Diego.

Profile 17. Financing Programs: Oklahoma and Utah The capital cost of conversion is, at present, a major economic hindrance to the commercialization of natural gas vehicle technology. (Once assembly line production of dedicated natural gas vehicles is firmly in place, the low price of natural gas will make natural gas vehicles much more cost-effective to own and operate than gasoline-powered vehicles.) Conversion kits cost between $2500 and $3000 for automobiles, including installation; between $5000 and $10,000 for medium-duty engines; and more than $20,000 for large urban transit buses. Financing for conversion projects is another incentive governments can provide to help reduce the barriers to natural gas use posed by these high capital costs. By increasing the number of vehicle owners who can afford to convert to natural gas use, government low-interest or no-interest loans can encourage development of a stronger market for natural gas vehicle technology. Some 13 states now have such programs.

Profile 17 describes financing programs in Oklahoma and Utah, the first two states to have such programs in place for alternative fuel vehicles.

Oklahoma's $1.5 million revolving loan fund offers seven-year, no-interest loans to state, municipal, and county government agencies and to school districts; the loans can amount to up to $3500 per vehicle converted and up to $100,000 per refueling station constructed. Although the loans can be made for conversions to any alternative fuel, applicants must demonstrate the ability to repay the loans from fuel savings; thus, they are largely restricted to natural gas, the only alternative fuel that is consistently cheaper than gasoline. As of October 1992, $1.2 million had been loaned to convert more than 150 vehicles, and fuel savings of more than $125,000 had already been achieved. The financing program (along with a tax credit program discussed in Profile 18) had attracted business to Oklahoma for vehicle conversions and construction of refueling stations.

Profile 18. Preferential Tax Treatment: Oklahoma and California Preferential tax treatment (including tax credits, tax exemptions, rebates, and lower tax rates) is another tool available to governments wishing to provide economic assistance to purchasers and operators of natural gas vehicles. Worldwide, each of these types of incentives has been used to spur investment in natural gas vehicle technology: in New Zealand and Canada, for example.

Profile 18 describes tax credit initiatives in Oklahoma and California, the states with the largest such programs for natural gas vehicle use. Oklahoma offers a tax credit of 50 percent of the cost of conversion equipment or refueling station equipment (decreasing to 20 percent in 1993). California offers two types of preferential tax treatment. Under one law, purchasers of vehicles that meet designated low-emission standards or of alternative fuel conversion equipment are exempt from paying sales tax on the incremental cost of this equipment compared to the cost of conventional vehicles. The second statute offers purchasers of these vehicles or conversion equipment tax credits equal to 55 percent of the incremental cost. Profile 18 also discusses tax credit programs offered by Connecticut, Colorado, New York, and the federal government, and ways of adapting road taxes based on gasoline sales to natural gas vehicles.

Profile 19. Favorable Regulatory Climate for Natural Gas Vehicle Investments: California Public Utility Commission Natural gas is among the most heavily regulated of all commodities sold in the United States, and the posture of the Federal Energy Regulatory Commission and state public utility or public service commissions with respect to the use of natural gas as a transportation fuel has a tremendous bearing on the future of natural gas vehicle technology. The key issue in the context of regulation of natural gas as a transportation fuel involves when investment in natural gas vehicle technology becomes a private expense (of the vehicle owner, or refueling station operator), rather than an expense borne by the local natural gas utility. In addition to determining which capital investments can be included in the overall rate base and thus charged to all natural gas customers, utility commissions have other regulatory tools with which they can either promote or hinder the development of natural gas vehicle technology: pricing, for example.

Profile 19 discusses California's regulatory actions because, perhaps more than any other state, California has grappled with the issues of utility commission regulation of natural gas vehicles and, in doing so, has provided effective encouragement for use of these vehicles. Responding to applications from three utility companies for rate-setting decisions regarding proposed major natural gas vehicle programs (San Diego Gas and Electric Company, Pacific Gas and Electric Company, and Southern California Gas Company), the California Public Utility Commission has allowed the companies to include program expenses in their rate base, offer rebates for conversions, and charge "incentive rates" for natural gas as a transportation fuel that are lower than the cost of producing and delivering the fuel.

Removal of Institutional and Regulatory Barriers

Despite the technological revolution in automotive transportation that is taking place, significant institutional barriers to the advancement of alternative fuel technologies in general, and to natural gas vehicle research, development, and use in particular, do exist. In order for natural gas vehicle technology to become rooted in United States culture as an accepted heir to gasoline-powered vehicles, public awareness and understanding must be greatly expanded. A comprehensive regulatory framework must be established that is specifically tailored to natural gas vehicles and that complements current laws and regulations affecting conventional vehicle use. Answers to questions about natural gas vehicles that are unanswered today must be sought, found, and incorporated into the regulatory framework. And all aspects of the new technology need to be monitored and reported, over time, to worldwide audiences. The final six profiles describe actions that are aimed at removing the regulatory and institutional barriers that currently hamper the widespread use of natural gas vehicles.

Profile 20. Setting Emissions and Safety Standards: US Environmental Protection Agency and the California Air Resources Board Dozens of federal and state laws and thousands of pages of regulations affect automotive production and use in the United States. To a great extent, these can be applied to natural gas vehicles as well as to gasoline-powered vehicles. In two crucial areas, however, natural gas vehicle technology is different enough to require special regulatory treatment: environmental and safety controls.

Profile 20 focuses first on emissions standards: those that the US Environmental Protection Agency is developing in response to the 1990 Clean Air Act Amendments and those adopted by the California Air Resources Board in 1990. The new EPA standards call for stringent limits on tailpipe emissions (a 60 percent reduction in nitrogen oxide and a 39 percent reduction in hydrocarbons, compared to 1992 model year cars) to be phased in starting with 1994 model year vehicles. These new emissions standards are likely to increase the economic competitiveness of natural gas vehicles, especially if the new regulations take into account the differing levels of problems caused by different types of hydrocarbons. The California regulations set even more stringent limits on emissions, establishing four stages of reductions in hydrocarbon emissions culminating in the introduction of so-called "zero-emission vehicles" by the turn of the century. If fully implemented, the California standards will constitute nothing less than the abandonment of the gasoline-powered vehicle in that state.

Profile 20 also discusses the need to set proper safety standards for natural gas storage cylinders and construction requirements for the design and operation of natural gas refueling stations, and current Department of Transportation and National Fire Protection Association efforts to do so. Finally, since all natural gas is not the same, the profile examines the need to set standards for natural gas quality; that is, the percentage of the gas that is methane and the percentage that is other, mostly inert, materials.

Profile 21. Removing Local Natural Gas Vehicle Use Restrictions: New York City A major task if natural gas vehicles are to gain widespread public acceptance is the identification, examination, and revision of local regulations that impede natural gas vehicle use. Nearly all such regulations were promulgated decades ago and were not drafted with transportation uses of natural gas in mind. Safety precautions were aimed at large energy storage depots, for example, as opposed to the smaller quantities of gas taken from local distribution lines and compressed for sale at individual natural gas refueling stations. Removing such local obstacles involves analyzing the true risks posed by natural gas vehicle use, especially in comparison to the risks posed by conventional vehicle use, and establishing local political support for the change.

Profile 21 describes a New York City controversy that offers a textbook example of how well-intentioned city leaders can unintentionally create regulatory compliance problems for natural gas vehicle owners and operators, and how such problems can be overcome. Following a 1949 accident involving a tanker truck loaded with a highly flammable substance in the Holland Tunnel joining New York and New Jersey, the Port Authority of New York and New Jersey (the agency responsible for the tunnel) banned vehicles carrying a wide range of hazardous materials, including compressed natural gas, from its tunnels and from the lower levels of bridges; the regulations were subsequently expanded to include other tunnels and bridges within the city itself. Thus, when natural gas vehicles were introduced years later, their routes into or out of the city were severely restricted. To overcome this prohibition, a group of organizations supporting natural gas use sponsored a risk assessment of the actual hazards posed by natural gas vehicles in tunnels; the study concluded that the overall risk was considerably less for natural gas vehicles than for gasoline vehicles under the most extreme hazard category examined. Convinced by the study, the Port Authority and New York City revised their regulations, leaving natural gas vehicles free to traverse the bridges and tunnels.

Profile 21 also discusses another case of restrictive regulation — affecting rooftop fuel storage — which illustrates the importance of evaluating the original intention of existing regulations as well as actual risk information when developing regulations for natural gas vehicles.

Profile 22. Investigating Unanswered Environmental Issues: The International Energy Agency The environmental edge that natural gas enjoys compared to gasoline or diesel fuel is a major reason for its appeal as a transportation fuel. However, the discovery of previously undetected flaws is not uncommon in emerging energy industries. Thus, natural gas vehicles cannot secure a central place in the United States transportation future unless unanswered environmental issues are investigated, assuring the public that natural gas use and natural gas vehicle technology do not contain some fatal, but as yet undetected, flaws. One area of potential concern is the emission of so-called "greenhouse gases," unregulated as of mid-1992, that are implicated in global climate change: carbon dioxide and methane.

Profile 22 examines the relative contribution of methane to global climate change and describes a multiyear International Energy Agency study comparing and estimating greenhouse gas emissions caused by the production and use of conventional automotive fuels and a variety of alternative transportation fuels. The study concluded that, when the entire fuel cycle from production to transportation to use is considered, natural gas vehicles will contribute 14.6 percent less to global warming than gasoline-powered vehicles, even though methane emissions from the average natural gas vehicle tailpipe will exceed those from the average gasoline-powered vehicle.

Profile 23. Providing a Knowledgeable Voice for Natural Gas Vehicle Interests: The Natural Gas Vehicle Coalition Profiles 20 and 21 examined the importance of changing federal, state, and local laws and regulations to make them more conducive to natural gas vehicle use. If this is to occur, those with an interest in the development and use of natural gas vehicles (from natural gas and automotive companies and natural gas vehicle owners and operators to environmental advocates and government officials seeking cleaner and less costly fuels) must organize to represent the case for natural gas vehicles in a clear and intelligent way. Although slow to develop, a unified voice advocating the value of natural gas vehicles is now emerging.

Profile 23 describes the first organization formed solely to promote the use of natural gas as a transportation fuel: the Natural Gas Vehicle Coalition.

Begun in 1988, its roots are firmly in the natural gas utility industry, but its membership has expanded to include natural gas vehicle equipment suppliers, state governments, trade associations, and educational institutions; there are also membership categories for nonprofit groups (such as environmental organizations) and individuals. NGVC work focuses on four areas: government affairs, technology, market development, and public relations and communications. The profile also highlights several other advocacy groups that are forming in both the public and private sectors, as well as a first-of-its-kind collaboration involving natural gas utilities and environmental organizations.

Profile 24. Promoting Public Education about Natural Gas Vehicles: Specialized Conferences and Publications One barrier to the development and commercialization of natural gas vehicles that can be easily overcome is the barrier of insufficient public awareness and understanding. The general public, energy policymakers, and technical specialists are often uninformed of the importance and need for the use of alternative fuel technology and the suitability of natural gas as an heir to gasoline as a transportation fuel. Until the considerable advantages of natural gas become widely known, natural gas vehicles are not likely to be in great demand.

Profile 24 describes the interest in natural gas vehicle information. It discusses the increasing number and popularity of specialized conferences and seminars; the role that specialized publications addressing alternative transportation fuels play; and the effectiveness of special media events for increasing the public visibility of natural gas vehicles.

Profile 25. Supporting International Information Exchange and Cooperation: The International Association for Natural Gas Vehicles Natural gas vehicle technology addresses energy and environmental concerns that are global in nature. Most countries get their oil from a small, politically volatile section of the world; thus, there is an international movement to diversify energy sources. At the same time, the implications of fuel use for the greenhouse warming of the planet attest to the global impact of transportation fuel emissions. Many countries are already undertaking natural gas vehicle programs; in order for the technology to develop efficiently, it is vital to have international information exchange and cooperation.

Profile 25 discusses the work of the International Association for Natural Gas Vehicles, the only organization that exists to serve global natural gas vehicle interests. Founded in 1986, the IANGV now has 180 members from 30 countries. The organization's objective is to increase natural gas consumption by the world's transportation systems; to this end, it conducts programs that highlight the efficiency, safety, durability, and environmental and economic advantages of natural gas vehicles. For example, it sponsors annual natural gas vehicle conferences; publishes a quarterly newsletter and annual yearbook; surveys natural gas vehicle programs worldwide; and participates in cooperative efforts to develop standard equipment specifications for critical natural gas vehicle and refueling station components.

Chapter 3 Natural Gas Vehicle Research, Development, and Demonstration

In 1876, the German inventor Nikolaus Otto produced a prototype of the internal combustion automotive engine still in use today. However, the use of oil-derived fuels to power the automobile engine was not a predetermined conclusion. Many early automotive engines were powered by other substances, such as gaseous fuels (including natural gas, propane, and hydrogen), coal, and even manure. Electric cars were popular in the 1890s; at the turn of the century, a French-built electric vehicle — the "Jenatzy" — held the world land speed record of 65.75 miles per hour.

Despite these initial experiments, the discovery of vast oil fields in the United States and the Middle East at the turn of the century lured the nascent automotive industry into a long-lasting dependence on oil-derived fuels, mainly gasoline and diesel. Although difficult and costly to refine, oil-based fuels were also attractive because their liquid state simplified refueling and fuel storage, and because their high energy density (a gallon of gasoline contains twice the energy as the same volume of methanol and four times the energy of compressed natural gas) minimized the space required by the fuel tank. In 1903, most of the 61,927 cars produced worldwide (almost 20 percent, or 11,235 vehicles, in the United States) were gasoline-powered.

Mass production of Henry Ford's Model T beginning in 1908 made reliable, standardized vehicles widely available for the first time; it also signaled the emergence of the United States as the world's leading automotive manufacturer and secured oil's position as the sole energy source for automobiles. Between 1908 and 1927, when Model T production ceased, 15 million of these cars had been sold at an average price of $440. Over the course of this century, the United States automotive industry has produced 500 million assembly line-built vehicles, virtually all of them fueled by gasoline or diesel.

The oil-fired, gasoline-burning internal combustion automobile engine stands among the most impressive technical achievements in modern history, and every year its design and performance are further refined and improved. Using one gallon of fuel, a typical internal combustion engine can, with comparatively little effort, haul two tons of machinery, people, and goods about 30 miles. Moreover, internal combustion engines can carry these loads gallon after gallon, day after day, until they have traveled the equivalent of several times around the world. In short, cars are durable, reliable, and inexpensive to operate.

In the 1990s, the world is facing a new technological challenge comparable to that faced by Henry Ford when he set out to standardize automobile production. The challenge is to develop and refine the technologies required to use fuels that are cleaner and cheaper than gasoline and diesel, and that come from more widespread and politically secure regions of the world than oil does. Meeting this challenge will require producing engines and fuel storage systems that can burn natural gas — and, eventually, a zero-polluting fuel such as hydrogen — instead of oil-derived fuels.

For decades, automotive engines have been designed to burn a low-quality, liquid fuel. Existing automotive technology does not take advantage of the "premium" characteristics of natural gas that allow it to burn more cleanly and efficiently than gasoline. A fundamental task necessary to encourage widespread commercial availability of natural gas vehicles is, therefore, to optimize automotive engines to burn natural gas.

This chapter discusses nine areas in which research, development, and demonstration of natural gas vehicle technology are now occurring. The chapter looks at the range of initiatives being undertaken in each of these areas, profiling in detail some of the key actors in each area. It also provides suggestions for future developments.

Natural Gas Research, Development, and Demonstration

What Needs To Be Done	Key Actors
1. Basic natural gas vehicle engine development	Gas Research Institute
2. Design and assembly line production of natural gas vehicles	Big Three auto producers
3. Light-duty fleet demonstrations	Gas utilities, government agencies, and Disneyland
4. Medium-duty commercial fleet demonstrations	United Parcel Service and Federal Express
5. School bus demonstrations	Garland and Harbor Creek School Districts and California
6. Urban bus system demonstrations	Federal Transit Administration
7. Use of liquefied natural gas in vehicles	Houston Metro and Roadway Express
8. Alternative fuel storage technology	Institute of Gas Technology
9. Natural gas and hydrogen fuel mixtures	Denver Hythane Project

Profile 1 Basic Natural Gas Vehicle Engine Development: The Gas Research Institute

While almost all of the 190 million vehicles on United States roads are powered by internal combustion engines, these engines come in a wide range of sizes and designs, from the four-cylinder, spark-ignition gasoline engines used in compact cars to the heavy-duty, compression-ignition diesel engines used in transit buses and trucks. Modifying and perfecting each engine type for efficient natural gas use requires extensive engineering research and development.

Natural gas vehicle engine research is underway at dozens of locations around the country. The organization funding the largest overall effort to perfect natural gas engine technology is the Chicago-based Gas Research Institute (GRI). Founded in 1976, GRI is supported by 300 member companies, including the largest pipeline companies, natural gas producers, and investor-owned and municipal gas utilities. GRI had a 1992 budget of $212.9 million; the budget is raised by a surcharge of 1.51 cents per thousand cubic feet of natural gas sold by member companies. The 250 staff members in GRI's Chicago headquarters develop and review applications for all kinds of natural gas research and contract with third parties to conduct that research. Very little research is actually performed within GRI itself.

Since GRI provided its first funding for natural gas vehicle research in 1988, its involvement in this area has soared. In 1990, GRI committed $4.1 million to natural gas vehicle research, and by 1991, this funding rose by more than 70 percent, to $6.9 million. As of mid-1992, GRI's five-year plan called for another $30 million in natural gas vehicle research, including $7.3 million in 1992 and $8.0 million in 1993. Because many GRI-funded projects include money contributed by other parties, such as the organization contracted to do the work, the total research effort is larger than the portion under GRI sponsorship.

Between 1989 and 1992, GRI contracted out more than 20 natural gas vehicle research projects. These projects focused on the three principal engine types: heavy-, medium-, and light-duty engines. Heavy-duty engines, typically diesel-powered, are used in urban transit buses and large trucks such as semi-trailers. Medium-duty engines power most delivery trucks, commercial service vans, and school buses. Engines in this category can be manufactured to burn either gasoline or diesel, but not interchangeably; once built, an engine can burn only the fuel for which it was designed. Light-duty engines are typically used in conventional automobiles, pickup trucks, and small vans. They are virtually all powered by gasoline.

Heavy-Duty Natural Gas Vehicle Engine Development Projects

GRI funds the development of natural gas vehicle technology for use in many of the most widely used types of heavy-duty engines in the United States; as of mid-1992, major projects included those described here.

- Several research projects focus on the engine that powers over 90 percent of United States urban transit buses: the 6V-92, nine-liter diesel engine manufactured by the Detroit Diesel Corporation (DDC). In one project, Stewart & Stevenson Company, the principal distributor of the engine, modified DDC's 6V-92 engine to burn diesel fuel and natural gas simultaneously. This dual-fuel engine is being tested in five buses operated by the Denver Regional Transit District. In another GRI-funded project, DDC itself is developing a dedicated engine that burns only natural gas.
- The Cummins Engine Company (of Columbus, Indiana) manufactures a 10-liter, L-10 diesel engine used in some urban buses and many heavy-duty trucks. Since 1988, it has been developing a dedicated natural gas version of this engine. By mid-1992, about 100 of these engines were being tested in the United States and Canada, primarily in urban buses. In August 1992, the L-10 engine was certified by the California Air Resources Board as meeting the state's stringent air pollution standards for heavy-duty engines. Full commercialization is expected in 1993.

- The Mack Truck Company's 12-liter diesel engine powers about 50 percent of all heavy-duty refuse haulers in the United States. This engine is being modified to burn natural gas using lean-burn technology (that is, technology that allows the engine to be flooded with excess air, thereby using less fuel and achieving greater fuel efficiency). It will begin to be tested near Mack Truck's headquarters in Boston in 1993. Commercialization is scheduled for 1995.

Medium-Duty Natural Gas Vehicle Engine Development Projects

Medium-duty vehicles are ideally suited for natural gas use because they are used almost solely by commercial fleet operators which are large fuel consumers, often equipped with centralized refueling stations. Not surprisingly, therefore, the largest number of GRI-funded research projects involve research and development of medium-duty natural gas vehicle engine technology, including those described here.

- Tecogen, a Massachusetts-based distributor of General Motors products, has modified a 427-cubic-inch (7-liter) GM gasoline engine to burn natural gas. Testing of the engine, called the TecoDrive 7000, in 10 California school buses began in 1991. In early 1992, GRI contracted with GM to refine this engine further and to bring it into commercial production. In mid-1992 the California Energy Commission purchased 100 school buses powered by the TecoDrive 7000 natural gas engine.
- In a project begun in September 1989, the California-based Acurex Corporation modified a 4.3-liter GM gasoline engine to burn natural gas. Field testing of this engine began in August 1991 with its installation in a United Parcel Service (UPS) delivery truck in Los Angeles. By mid-1992, the field test program had been expanded to include UPS vehicles in five additional cities: Dallas, New York City,

Oklahoma City, Tulsa (Oklahoma), and Washington, DC.

Entenmann's Bakery has selected the Hercules natural gas engine for field testing in its delivery trucks. Photo: Gas Research Institute

- Hercules Engines, Inc., of Canton, Ohio, is modifying two of its medium-duty gasoline and diesel engines to burn natural gas: a four-cylinder, 3.7-liter engine, a type commonly used in panel van delivery trucks, and a six-cylinder, 5.6-liter engine often used in school buses. By 1993, field testing of the four-cylinder engine is scheduled to begin in three Entenmann's Bakery delivery vans. Entenmann's already operates 32 trucks with conventional gasoline engines retrofitted to burn natural gas.
- Illinois-based Navistar International Transportation Corporation's 7.3-liter diesel engine is used in about 50 percent of United States school buses. The company is directing another GRI-funded project, begun in 1991, to modify this engine for natural gas use. When commercialized in 1995, the engine will be fitted to a school bus chassis that is being designed by Navistar to accommodate the natural gas storage tanks.
- In August 1991, Cummins Engine Company began a GRI-funded project to modify its 359-cubic-inch (5.9-liter) 6BTA diesel engine so that it can burn natural gas. Commercial availability of school buses and delivery trucks equipped with the modified Cummins engines is expected in 1994.

Light-Duty Natural Gas Vehicle Engine Development Projects

In 1991, GRI for the first time entered into research contracts with each of the Big Three United States automotive producers (General Motors, Ford, and Chrysler) for development of conventional light-duty car and truck engines that will burn natural gas. GRI-funded projects to develop light-duty natural gas vehicle engines include those listed here.

- General Motors is developing a 5.7-liter engine for use in its Sierra pickup truck.
- Ford is developing a 4.6-liter engine for use in its F150 pickup truck.
- Chrysler is developing a 5.2-liter engine for use in a full size "B-Van."

All of these light-duty vehicle projects are discussed in greater detail in Profile 2.

Additional Gas Research Institute Projects

In addition to natural gas vehicle engine research and development work, GRI is funding projects to improve the technology for natural gas storage in vehicles. This work is described in Profile 3.

GRI also frequently publishes research reports that examine issues related to natural gas vehicle technology. Two 1991 publications of this type were *Potential for Compressed Natural Gas Vehicles in Centrally-Fueled Automobile, Truck and Bus Fleet Applications* and *A White Paper: Preliminary Assessment of LNG Vehicle Technology, Economics, and Safety Issues.*

GRI has played a critical role in "jump starting" research and development of optimized natural gas vehicle technology by the private sector. Through the contracts it has awarded, several dozen companies have initiated or expanded natural gas vehicle programs. The positive results of these GRI-funded projects have prodded other companies to enter the natural gas vehicle field. As a result, virtually every automotive or truck engine manufacturer is now developing a natural gas vehicle product line for commercialization in the mid-1990s.

Profile 2 Design and Assembly Line Production of Natural Gas Vehicles: The Big Three Auto Producers

The Big Three United States auto producers (General Motors, Ford, and Chrysler) annually produce about one-third of the world's new cars, including conventional passenger sedans and light-duty pickups, wagons, and vans. They manufacture about 92 percent of the cars built in the United States, and they employ most of the 676,000 workers at this country's 335 automotive manufacturing plants.

In 1991, United States consumers paid approximately $100 billion for the 6.9 million cars built by the Big Three. None were natural gas vehicles. Until 1992, natural gas vehicles were created by retrofitting conventional gasoline-powered cars with equipment capable of storing and burning natural gas. This retrofit procedure is termed "aftermarket" conversion.

A serious commitment by the Big Three to assembly line production of natural gas vehicles is essential if natural gas vehicles are to become a major component of the United States automotive market. In undertaking large-scale natural gas vehicle production, the Big Three have several advantages unavailable to smaller manufacturers. First, their production systems can achieve economies of scale unavailable to smaller producers. Second, the Big Three have established distribution, marketing, and advertising networks that can also be used to build a market for natural gas vehicles. Third, because they produce and therefore can offer warranties for the entire vehicle, they can also promote consumer confidence by selling fully warrantied natural gas vehicles. Finally, building vehicles to burn natural gas in the first place eliminates the additional labor needed to modify an existing vehicle through an aftermarket conversion.

The reluctance of the Big Three to embrace alternative transportation fuels, including natural gas, has historically been one of the largest obstacles to alternative fuel use. Although the world supply of natural gas vehicles grew to about 500,000 cars in the 1980s, virtually all were converted gasoline-burning vehicles; the Big Three converted fewer than 100 natural gas vehicles during that period.

Effect of the 1990 Clean Air Act Amendments

Changes in federal and state laws have spurred the Big Three to take an active interest in the design and production of vehicles using a variety of alternative fuels, including natural gas. The 1990 Clean Air Act Amendments, for example, have created a demand for up to one million alternatively fueled vehicles by the turn of the century through the requirement that commercial fleet vehicles in the cities experiencing high ozone air pollution burn "clean" fuels.

In the short term, the automotive industry's major response to the 1990 Clean Air Act Amendments has been to study, in collaboration with the major oil companies, the potential for using reformulated gasolines to comply with the emission requirements that will go into effect in the next few years. To address the longer term requirements of the law, however, the industry has begun experimenting with various alternative fuels. For instance, each of the Big Three has developed prototype electric vehicles, and in 1991 they jointly established a four-year, $260 million project called the United States Advanced Battery Consortium to develop batteries suitable for automotive use. Moreover, in the next few years, the Big Three will be producing several thousand methanol vehicles for sale in California; full production runs of methanol vehicles may follow.

In addition, since 1990, each of the Big Three has begun designing and producing natural gas vehicles, planning to commercialize them by 1995. The Big Three's investment in natural gas vehicle technology, involving tens of millions of research dollars since 1990, has probably surpassed the combined expenditures in all of its alternative fuel programs during the 1980s.

General Motors

As of mid-1992, General Motors had the largest natural gas vehicle program of the Big Three. GM is offering an "early entry" natural gas vehicle; that is, a commercially available natural gas vehicle designed and built to burn natural gas, and produced in small numbers. By 1995, GM plans to introduce several full-production vehicle lines dedicated specifically to burning natural gas. In 1991, GM became the first of the Big Three to join the Natural Gas Vehicle Coalition, the leading trade organization for groups interested in natural gas vehicles.

GM's early entry program was unveiled in July 1990, following negotiations between GM and a ten-company consortium of natural gas utilities located in Colorado, California, and Texas. These companies agreed to purchase GM's first 1000 natural gas vehicles. GM selected its 3/4-ton Sierra pickup truck with a 5.7-liter, 175 horsepower engine as the prototype; first deliveries of the Sierra natural gas vehicle began in late 1991.

Production of GM's Sierra natural gas vehicle begins by taking trucks that have been manufactured as gasoline-burning vehicles at a GM factory in Pontiac, Michigan. The trucks are then shipped to PAS, Inc., in Troy Michigan, where they are transformed, according to GM's specifications, into dedicated natural gas vehicles. PAS removes the gasoline tanks and replaces them with three aluminum-lined composite natural gas storage cylinders. The tanks carry enough natural gas for 180 miles of driving between refuelings. The aluminum tanks add only 156 pounds to the vehicle, compared to the conventional gasoline storage system, not enough to significantly reduce the vehicles' power or performance. The natural gas Sierra accelerates from 0 to 60 miles per hour in 10.6 seconds, just 0.4 seconds off the pace of the gasoline model. The Sierra natural gas vehicles are also equipped with automatic transmissions and closed-loop, electronically controlled air pollution control equipment which has been certified as meeting California's strict 1992 emission standards.

By mid-1992, about 160 natural gas-fueled Sierra trucks were being built per week, and production was committed at least through 1992. In fact, according to GM, orders for the Sierra far exceeded the initial target of 1000. Consequently, in early 1992, GM announced that as many as 10,000 natural gas-powered Sierra trucks will be built. Each vehicle costs about $3700 more than a similar model powered by gasoline.

The early entry project was barely underway when negotiations began concerning an expanded effort to design other GM engine and vehicle types to burn natural gas. This effort involves four engine types and up to a dozen light-duty and medium-duty truck models. The cost of this second phase of the research and development program is estimated to be $40 million, of which natural gas utilities will contribute $16 million and GM will contribute the remaining $24 million. The schedule calls for commercialization of natural gas vehicles developed under this program by the mid-1990s.

Chrysler

Like GM, Chrysler has begun production of a dedicated natural gas vehicle for commercial sale. Since its program began in 1991, Chrysler has invested more than $3 million in natural gas vehicle research and development. Chrysler's development program began with the production and testing of 25 prototype vehicles by the end of 1991. In March 1991, furthermore, Chrysler won a contract to supply the federal General Services Administration with 50 natural gas-powered, eight-passenger B-vans over a one-year period. These vehicles, equipped with a 5.2-liter engine and three natural gas storage cylinders, offer a driving range of about 120 miles.

The B-vans delivered to the federal government were the first of a much larger production effort anticipated by Chrysler. In March 1992, Chrysler began taking orders from other parties for what it anticipated to be a 1992 production of 2000 B-vans. Conventional B-vans, being built at the company's Pillette Road Truck Plant in Windsor, Canada, are converted on-site into dedicated natural gas vehicles.

Ford

Ford Motor Company has experimented with natural gas technologies longer than its competitors, but its current program is the most cautious among the Big Three. Over the last 40 years, Ford has produced and tested 1244 demonstration vehicles of various types. The company has experimented with a wide variety of alternative transportation fuels, including natural gas, methanol, ethanol, propane, and electricity.

In 1984, Ford converted 27 Ranger pickup trucks to burn only natural gas. The vehicles were leased to gas utility companies and tested for five years. Although their performance matched that of conventional gasoline-powered Rangers, Ford dropped the program because it received no large purchase orders as a result of the demonstration, probably because oil prices were in the middle of a steep decline.

In 1990, as renewed interest in alternative transportation fuels emerged, Ford reestablished its natural gas vehicle program to develop a dedicated natural gas-burning 5.0-liter engine for use in pickup trucks (F150 series). A 1991 feasibility study included the building and testing of ten prototypes. In 1992, Ford built about 100 natural gas-powered pickups, and it projects that an additional 200 to 600 will be produced in 1993. Future development phases will include final engine and chassis design and testing prior to full-scale production, probably in 1994.

A second Ford program involves the production of a Crown Victoria sedan dedicated to burning natural gas. In April 1992, the company delivered five Crown Victoria natural gas vehicles, each equipped with a 4.6-liter, V-8 engine, to Southern California Gas Company for use in a demonstration program. A total of 50 natural gas-powered Crown Victorias were built in 1992 and leased to about 25 utility companies around the United States and Canada for demonstration purposes.

Profile 3 Light-Duty Fleet Demonstrations: Gas Utilities and Disneyland

Of the millions of vehicles on the road in the United States today, some are more suitable for conversion to natural gas than others. The family station wagon, for example, which sits in a garage much of the time but is called upon once a year to take the kids and the dog across the country on a camping trek to the Grand Canyon, is one of the least suitable. Its use pattern — limited driving, long-distance outings through sparsely populated regions, and great demand for cargo space — makes it unable to overcome the major shortcomings of current natural gas vehicle technology.

Conversely, commercial fleet vehicles are ideally suited for demonstrating the economic and other advantages of natural gas vehicle use. Most fleet vehicles are in constant use. As a result, the operator quickly experiences the key economic benefit of using natural gas: its typically low price compared to gasoline. The fact that present natural gas vehicle technology limits most natural gas vehicles to a geographic range of about 200 miles between refuelings is not a problem for most fleets, especially those in urban areas, because most commercial fleets are driven less than that distance daily, along predictable routes such as daily delivery or service call stops. In addition, a majority of fleet vehicles return each day to a central location for refueling overnight, thus avoiding the need for a natural gas refueling infrastructure. Furthermore, the need for cargo space is less critical since a large percentage of light-duty fleet vehicles mostly carry employees rather than loads. Finally, because fleet vehicle demonstrations can involve relatively large numbers of vehicles, they help prove that natural gas can be successfully used as a fuel on a wide scale.

In the United States today, there are approximately 11 million passenger cars and light-duty trucks in fleets of 10 or more. Each year, 1.7 million new fleet vehicles are purchased. Private industry fleets comprise the largest sector of the fleet vehicle market, with more than 7.9 million vehicles in fleets of 10 or more. Recent American Gas Association studies predict that more than one million and as many as six million alternative fuel light-duty automobiles and trucks could be operating in United States fleets shortly after the turn of the century as a result of aggressive implementation of commercial fleet clean fuel programs.

Although the future looks promising for natural gas vehicles, with hundreds of natural gas vehicle programs underway involving private and public sector automotive fleets, fewer than 1 percent of United States fleet vehicles in all categories now run on natural gas. Even the natural gas industry, an obvious and major natural gas vehicle market, has converted fewer than 5 percent of its vehicle fleet. Clearly, much work is needed before natural gas vehicle use is widespread in commercial fleets.

This profile discusses two examples of one fleet market that could successfully implement natural gas vehicle technology on a wide scale: light-duty vehicles. The following three profiles discuss different fleet markets with similar potential: medium-duty trucks, school buses, and urban transit buses.

Utility Projects

Not surprisingly, natural gas companies were among the first to demonstrate the advantages of natural gas vehicle technology in their own fleets. An American Gas Association survey completed in October 1991 revealed that 8827 natural gas vehicles are part of fleets operated by 82 United States natural gas utilities. Nearly all of these natural gas vehicles are bi-fuel automobiles and service vehicles, capable of burning either natural gas or gasoline.

The Northern Indiana Public Service Company (NIPSCO) in Gary, Indiana conducts one of the country's largest utility natural gas vehicle programs. The NIPSCO program started in 1981, and by 1986 the company's natural gas vehicle fleet peaked at 938 natural gas vehicles. Corporate restructuring reduced NIPSCO's natural gas vehicle fleet to 640 (out of 3100) vehicles as of mid-1992. Of the total, 615 vehicles are bi-fuel, capable of burning either natural gas or gasoline; the remaining 25 are dedicated to natural gas. The bi-fuel vehicles run approximately 80 percent of the time on natural gas and they generally switch to gasoline only after the natural gas supply is exhausted. NIPSCO reports savings of about $250,000 in annual fuel costs by using natural gas instead of gasoline.

NIPSCO plans to purchase several dedicated natural gas GM Sierra pickup trucks by the end of 1992, and perhaps a B-van. It projects the total size of the natural gas vehicle fleet to rise to 775 vehicles in 1993, including 125 dedicated natural gas models. By 1995, the company expects to have more than 1000 natural gas vehicles on the road.

Disneyland

Since 1968, when a double-decker bus operating along Main Street became the first Disneyland vehicle to be converted to natural gas, natural gas vehicles have constituted the backbone of Disneyland's transportation system. Disneyland switched to natural gas to reduce air pollution, eliminate water clouding from boat emissions at marine rides, and reduce fuel bills.

Maintenance vehicles at Disneyland in California run on natural gas. Photo: © The Walt Disney Company

Disneyland's current natural gas vehicle fleet includes 60 boats and rafts, one submarine, four Main Street vehicles, and 16 back-lot service vehicles. After decades of operation, there has never been an accident or injury at Disneyland resulting from the operation of a natural gas vehicle. The natural gas-powered rides have operated as safely as gasoline-powered rides, at less cost and producing less pollution. The current natural gas vehicle fleet saves $56,000 a year in fuel costs compared to a comparable fleet operating on gasoline. Disneyland's long-term demonstration program thus provides an encouraging example for other operators: it proves that Tomorrowland can be here today.

Profile 4 Medium-Duty Commercial Fleet Demonstrations: United Parcel Service and Federal Express

Nearly a quarter of the 190 million vehicles in the United States, 43 million, are trucks, and truck sales amount to more than 4.0 million vehicles per year. About 64 percent of these trucks, or 27.5 million, are panel, pickup, and small service trucks weighing less than 6000 pounds. Medium-duty freight trucks, such as commercial delivery vans, weighing between 6000 and 10,000 pounds, comprise another 24 percent (10.3 million). Heavy-duty vehicles, such as refuse trucks, semi-trailers, and buses, weighing up to 30,000 pounds, account for the remaining 12 percent (5.2 million).

From an economic standpoint, trucks present an especially suitable market for natural gas vehicle development and demonstration, because their high fuel consumption rates make them ideal candidates for conversion to natural gas use. Out of all vehicle categories, trucks are, on average, annually driven the most miles per vehicle, about 13,400 miles per year, compared to 9600 for the average passenger vehicle. A typical truck gulps 1359 gallons of fuel yearly, nearly three times the 500 gallons consumed by the average car. A semi-trailer truck driven once across the country consumes more fuel than an average automobile burns in one year.

The low cost of natural gas compared to gasoline or diesel rewards users of natural gas-powered trucks with fuel savings that can offset the capital costs associated with converting vehicles to natural gas use in a matter of a couple of years — resulting in significant bottom-line savings for businesses over the lifetime of the vehicle.

Natural gas-powered trucks can also play a vital role in ensuring energy security, since trucks deliver most products to United States markets. In the event that imported oil supplies are interrupted, natural gas-powered trucks can continue to provide vital services, such as food and goods delivery.

Further, the driving pattern of commercial trucks can easily and economically be accommodated to the current state of natural gas vehicle technology. More than 70 percent of commercial trucks are returned to a central location each night, and 85 percent of these vehicles typically travel less than 200 miles during an average business day. Medium-duty commercial trucks, especially those now powered by gasoline engines, are thus beginning to receive significant attention for commercial natural gas vehicle demonstration programs.

United Parcel Service Demonstration Program

A United Parcel Service package car, converted to run on compressed natural gas, is refueled at a "Quick Fill" station. Photo: UPS

As of mid-1992, United Parcel Service (UPS) was conducting the most ambitious medium-duty commercial fleet natural gas vehicle demonstration program in the United States. The importance of the UPS demonstration program results from the delivery service's size. Each day, 119,000 brown and gold UPS trucks deliver 11.5 million packages worldwide, driving 5.5 million miles in the process. The company's 244,000 employees work in 1766 facilities in more than 180 countries; it operates the world's ninth largest airline of over 400 planes.

In August 1989, UPS took the first step in a multiphase natural gas vehicle demonstration program. Ten UPS delivery trucks were converted to run on natural gas in New York City. The natural gas vehicles were tested with ten comparable gasoline-fueled trucks. The two groups of test vehicles drove more than 70,000 miles on the streets of New York City during the program's first year. The converted UPS trucks, which retained the capability to burn gasoline, ran on natural gas for more than 90 percent of the test period. Brooklyn Union Gas Company contributed a natural gas refueling station for the trucks.

After one year, the test results favored continued use of natural gas. Daily maintenance expenses for the natural gas vehicles were about $1.00 lower per vehicle and the natural gas itself cost about 30 percent less than gasoline on an equivalent energy basis; that is, the amount of natural gas that yields the same amount of energy as a gallon of gasoline costs 30 percent less. Moreover, the natural gas vehicles were 12 percent more fuel efficient than gasoline-powered trucks: using the same amount of energy they could run 12 percent further. In addition, the operating performance of the two types was virtually indistinguishable with regard to power, acceleration, and safety.

The UPS natural gas program also demonstrated significant reductions in pollution. Compared with emissions from the gasoline-powered vehicles, emissions from the natural gas vehicles of non-methane hydrocarbons and nitrogen oxides, the key ingredients in smog, were 86 percent and 22 percent lower, respectively. Carbon monoxide emissions were 18 percent lower, and evaporative emissions were virtually eliminated in the natural gas vehicles.

Encouraged by its experience with natural gas vehicles in New York City (where the program is continuing), UPS has broadened the demonstration program to include five other cities: Washington, DC; Dallas; Los Angeles; Tulsa; and Oklahoma City. The Washington program, begun in the fall of 1991, involves three natural gas-powered trucks that make daily deliveries to the Capitol building and are refueled at a nearby station constructed jointly by Amoco and Washington Gas Company. In Dallas, UPS began a two-year test program in October 1991 involving 15 delivery trucks and a refueling station built by Lone Star Gas Company.

In California, UPS' Los Angeles program, also begun in 1991, involves 20 trucks and a refueling station built by Southern California Gas Company. If the program is successful during a two-year test period, UPS has announced its willingness to convert all of its 2700 trucks in the Los Angeles area to natural gas use, in part to help the region battle its severe air quality problems.

In January 1992, UPS unveiled its largest natural gas vehicle venture to date. Lured by a 50 percent state tax credit in Oklahoma, the company decided to convert 140 delivery trucks to natural gas at a cost of about $3500 each. Seventy trucks will operate in the state's two largest cities: Tulsa and

Oklahoma City. The company will also build a natural gas refueling station in each city at a cost of $250,000 each.

Federal Express

In October 1990, one of UPS' leading competitors, Federal Express, started its own alternative transportation fuels program. The Federal Express "CleanFleet" program involves 103 express package delivery trucks in Los Angeles that will be converted to use a variety of alternative fuels (natural gas, ethanol, methanol, propane, reformulated gasoline, and electricity) and tested over two years. The demonstration will include twenty vehicles converted to run on each of the first five fuels, and three electric vehicles.

The first step in the Federal Express project was a $450,000 project feasibility study conducted by Battelle Memorial Institute, a private research and engineering company based in Columbus, Ohio. Completed in early 1991, this report provided the basis for the development of a consortium of companies (including Federal Express) and government agencies that are funding the project. Total direct project costs are estimated to be $6.5 million, with an additional $4.2 million in in-kind services offered by project participants. The contribution of Federal Express will total approximately $1 million.

In addition to Federal Express, other corporate project participants include the Big Three auto manufacturers, Southern California Gas Company, ARCO Products Company, Chevron USA Inc., the American Methanol Institute, the Gas Processors Association, the LP Gas Clean Fuels Coalition, the National Propane Gas Association, Southern California Edison Company, and the Western Liquid Gas Association. Government participants include the California Energy Commission, the California Air Resources Board, the South Coast Air Quality Management District, the US Department of Energy, and the US Environmental Protection Agency.

Project CleanFleet was formally initiated in June 1992. As part of the natural gas vehicle component of this project, each of the Big Three auto manufacturers has provided about seven 1992-model natural gas vehicles to Federal Express. They will all be tested for two years in the Federal Express fleet in Irvine, California.

Profile 5 School Bus Demonstrations: Garland and Harbor Creek School Districts and California

School bus fleets offer a particularly promising opportunity for using and demonstrating the benefits of natural gas vehicles for at least six main reasons.

1. The market is large: 81 percent of all buses operating in the United States — or over 500,000 vehicles — are school buses. They carry more than 22 million children to and from school each day.
2. Given present natural gas vehicle technology, school bus driving patterns are ideally suited to natural gas use. Like the commercial light-duty and medium-duty commercial fleets discussed in Profiles 3 and 4, school buses can easily be refueled because the average school bus travels less than 200 miles each day and the vehicles are normally housed together in a central location.
3. The economics of operating school bus fleets favor natural gas use. Because of their size and weight, school buses achieve poor gas mileage. Thus, comparatively cheap natural gas presents attractive fuel cost savings for operators of school bus fleets.
4. Most school buses remain idle much of the day while school is in session, so refueling can often be scheduled conveniently, even during daytime hours.
5. Unlike most vehicles of this large size, many school buses are powered by gasoline engines which are easier to convert to natural gas with current technology than are diesel engines.
6. Because school buses typically have high road clearances, they provide ample room for safe installation onto the chassis of natural gas storage cylinders.

Safety of fuel use in school buses is of special concern because these buses are primarily used to transport children. Experience to date indicates that natural gas use in school buses is at least as safe and probably safer than gasoline use. (Safety issues were discussed in more detail in Chapter 1.) Moreover, installation of sensitive natural gas detectors (which sound an alarm when there is even a small leak of natural gas) inside the bus or on the underside of the chassis adds an additional level of safety to natural gas-fueled buses.

The American Gas Association estimates that at least 25 school districts around the country have converted five or more buses to natural gas, and this number is growing. The economic benefits of converting to natural gas are proving to be considerable. For example, with average annual fuel savings of $1500 per bus, converting a fleet of 30 school buses to natural gas use could generate enough savings in fuel costs to fund a full-time, salaried teaching position. Partly for this reason, in Texas, which operates 25,000 school buses (more than any other state except New York), a 1989 state law requires conversion of up to 90 percent of all school buses in fleets of more than 50 over the next decade. (See Profile 15 for a discussion of the Texas mandate.)

Garland, Texas Project

As of mid-1992, one of the largest school bus conversion programs undertaken had been conducted by the Garland Independent School District (GISD), northeast of Dallas. About 180 GISD buses transport 10,000 students daily to schools in this community. More than half of these buses, 93 as of early 1992, operate on natural gas.

Attracted by the low cost of natural gas, GISD began its program in 1982, long before the Texas law mandating conversion was in effect. GISD's demonstration program was implemented in three phases. Phase I, completed in 1985, saw the conversion to natural gas use of 65 buses. Three refueling stations were also constructed: one slow-fill operation capable of refueling 60 buses overnight and two fast-fill pumps, each of which can refuel a natural gas-powered bus in less than 10 minutes. (The technological differences and comparative advantages of slow-fill and fast-fill pumps are discussed in Profile 12.) Phase II, completed in 1986, involved conversion of 16 more buses. Phase III, completed in 1991, included about a dozen additional conversions and the expansion of refueling equipment. Total capital costs for the GISD natural gas bus program amount to $389,773: approximately $241,000 for Phase I and $150,000 for Phases II and III.

Capital costs have been more than recouped through fuel savings and reduced maintenance expenses. Fuel savings in the initial year of the program were $57,614. These have grown to over $100,000 annually, or about $1690 per bus operating on natural gas. GISD's natural gas program has resulted in other cost savings as well. Because of the clean-burning quality of natural gas, natural gas use has extended GISD's cycle for oil changes from 3000 to 8000 miles. Furthermore, in natural gas-powered buses, spark plug life has nearly doubled, exhaust system repairs have dropped 63 percent, and carburetor repairs have been cut 85 percent. As a result, total bus fleet maintenance costs have been reduced by more than $4000 per year.

Harbor Creek, Pennsylvania Project

Another pioneering project that continues to provide economic benefits involves the conversion of 40 buses by the Harbor Creek School District outside of Erie, Pennsylvania. Started in 1981, within 21 months the project had fully recovered the $147,000 in capital costs required to convert the buses and construct a refueling station. Harbor Creek reports that its natural gas-powered buses have been found to start more easily and more reliably on cold winter days. They also achieve better fuel efficiency than their gasoline-powered counterparts. Annual fuel savings from natural gas use now total about $50,000.

California's Demonstration Program

Nearly all school bus demonstration programs to date have converted gasoline-powered buses to burn either gasoline or natural gas. A 1988

California law, however, authorized creation of a program to test school bus engines designed specifically to burn only alternative transportation fuels. Under the $100 million "Safe School Bus Clean Fuel Efficiency Demonstration Program" conducted by the California Energy Commission, buses built prior to 1977 must be replaced, as they are retired, with new buses that burn clean alternative fuels.

The program's first 10 natural gas-powered buses, delivered in 1991, are equipped with a 427-cubic-inch General Motors engine, modified by Tecogen to burn natural gas. (The engine, called TecoDrive 7000, is the product of the research project funded in part by the Gas Research Institute discussed in Profile 1.) The bus chassis for the demonstration model was built by GM, and the body was produced by Blue Bird Body Company.

Up to 300 more natural gas-powered buses could eventually be included in the California demonstration program. In mid-1992, California ordered an additional 100 TecoDrive-powered, 84-passenger natural gas school buses as part of this program. Several other manufacturers of school bus engines, including Hercules and Navistar, are also developing dedicated natural gas engines that are likely to be incorporated into California demonstration projects during the next few years.

The Federal Effort

In mid-1992, the US Department of Energy funded 10 school bus demonstration projects as part of its first effort to encourage use of alternative transportation fuels in school buses. Total project funds of $750,000 will pay the incremental cost of alternative-fueled buses compared to the cost of conventional buses. The 10 grants followed an initial grant in early 1992 to the school district in Wood County, West Virginia, to pay for eight natural gas school buses. The 10 newer projects are located in Arizona, Maryland, Missouri, Kentucky, New Mexico, New York, Pennsylvania, Vermont, Utah, and the District of Columbia.

Profile 6

Urban Bus Demonstrations: The Federal Transit Administration Alternative Fuels Initiative

Most of the 1.0 million large, heavy-duty trucks and buses on the road today — including the buses typically used in urban mass transportation systems — are powered by diesel fuel. There are about 58,000 urban transit buses in the United States today, and about 3000 new ones are introduced each year. Of all heavy-duty vehicle types, urban buses have to date provided the largest market for demonstration of natural gas technology, in large part because of public pressure to eliminate the noxious dense black smoke emitted from the tailpipes of many diesel fuel-powered buses.

Urban buses present an ideal opportunity for natural gas conversion from an environmental standpoint. Diesel engines are energy efficient and powerful, but they are also egregious air polluters. An urban bus emits 500 times more particulate air pollution than the average passenger car. Because urban buses operate in highly populated areas, their pollution affects the health and quality of life of large numbers of people. The 1990 Clean Air Act Amendments require major reductions in particulate emissions from new buses beginning in 1993.

Natural gas-powered buses, by contrast, produce virtually no particulate pollution. Thus, the use of natural gas in urban buses directly reduces pollution from bus tailpipes. This has a beneficial side effect: by burning fuel more cleanly, natural gas-powered urban buses can improve the public image of mass transportation. In addition, because natural gas use reduces fuel costs, it can help lower fiscal pressures on financially strapped municipalities.

In 1988, Brooklyn Union Gas Company pioneered the demonstration of natural gas in urban buses by funding the design and construction of the nation's first two dedicated natural gas-burning buses. Built by Bus Industries of America, Inc., and operated by Command Bus Company in New York City, the buses were originally equipped with large gasoline-burning engines converted to run on natural gas. In 1992, these engines were replaced with dedicated natural gas-burning diesel engines, and they will continue to operate on natural gas into the future.

The Challenge of Converting Diesel Engines

Converting diesel engines to natural gas use has proven more difficult than converting gasoline engines: there are many technological hurdles to overcome before perfecting natural gas use in diesel engines. Diesel engines are desirable for heavy-duty transportation because they are durable, reliable, and energy-efficient. In part, their reliability and efficiency derive from the fact that diesel engines do not contain spark plugs, which play a key role in igniting the fuel used in conventional gasoline engines. By contrast, diesel fuel is ignited solely by the heat generated as a result of compressing the fuel. Natural gas and gasoline, however, do not ignite under the high compression that exists in diesel engines. Thus, the design of diesel engines requires modifications so that natural gas will burn. Three approaches to resolving this technological obstacle to the use of natural gas in heavy-duty diesel engines are successfully being applied in research and demonstration projects.

One effort involves dual-fuel technology: modifying the engine so that natural gas and diesel fuel are simultaneously injected into it. (Dual-fuel technology is to be distinguished from bi-fuel technology, which allows for the use of either gasoline or natural gas, but not both simultaneously.) Using dual-fuel technology, an engine is fired almost entirely with diesel fuel when it is first started and idling. As the engine warms up and is operated at high power, it continuously generates sufficient heat to ignite natural gas in the absence of diesel fuel. Dual-fuel engines can substitute natural gas for diesel fuel about 80 percent of the time.

Dual-fuel diesel technology was first developed in the mid-1980s in Canada. It was field tested in the United States in two buses operated by Pierce Transit Company in Tacoma, Washington. More recently, the Gas Research Institute has funded the development of a dual-fuel version of the Detroit Diesel Corporation 6V-92 engine, the most commonly used urban bus engine in the United States. (GRI's diesel research programs are discussed in Profile 1.)

The two other efforts to adapt diesel engines to natural gas use involve the introduction of a heat source into the diesel engine cylinder. One approach adds a spark ignition system to the engine, essentially converting the diesel engine into a standard spark-ignited (or Otto cycle) engine. The second approach adds a heat source to each engine cylinder in the form of a hot wire coil, called a glow plug, similar in form to an automobile cigarette lighter. Like spark plugs, glow plugs are hot enough to ignite natural gas in the absence of diesel fuel. Cummins Engine Company and the Detroit Diesel Corporation are developing diesel engine designs that incorporate both of these natural gas-burning modifications.

Federal Transit Administration Demonstration Project

In 1988, the United States Urban Mass Transportation Administration, renamed the Federal Transit Administration (FTA) by the November 1991 Intermodal Surface Transportation Efficiency Act, launched what, through mid-1992, has been the largest federal alternative transportation fuels demonstration program involving heavy-duty vehicles. Called the "Alternative Fuels Initiative," the program offers federal funds to cities for the purchase of urban buses that can either burn alternative fuels or are equipped with advanced air pollution controls to reduce diesel smoke. Under the initiative, FTA generally pays 90 percent of bus purchase costs.

The transit authority in Cleveland, Ohio, purchased natural gas buses with FTA funds. Photo: East Ohio Gas

As of mid-April 1992, 92 applications covering nearly 2000 buses powered by alternative transportation fuels were pending before FTA (derived from the more than 1300 applications submitted to FTA from cities in 26 states since the beginning of the initiative). At an average price of $200,000 for an urban transit bus modified to burn alternative fuels, this represented a possible investment of $400 million for buses powered by clean fuels. By mid-1992, more than 200 buses were on the road because of the Alternative Fuels Initiative, and FTA had approved the purchase of an additional 300 vehicles.

Natural gas projects predominate in the FTA program. Indeed, there are more approved orders for natural gas buses than for buses powered by all other alternative fuels combined. About half the buses already in operation are powered by natural gas. The second largest category consists of methanol-powered buses (primarily the 59 methanol buses already funded by FTA when the initiative began, but tabulated as part of the program anyway). Other projects include funding for ethanol-powered buses and buses fueled by liquefied petroleum gas.

The largest FTA grant as of mid-1992, totaling $6.2 million, was awarded to the Pierce Transit Company in 1991. The grant funded the purchase of 50 natural gas buses to supplement the two prototypes Pierce Transit built on its own initiative in the mid-1980s. Nineteen of the new buses are 28 feet long and can carry 24 passengers each. These buses are equipped with bi-fuel engines capable of burning either natural gas or gasoline. The other 31 buses are larger, 40-foot-long vehicles powered by the dedicated natural gas-burning L-10 engine developed by Cummins Engine Company. The project also includes the construction of one of the largest natural gas refueling stations in the United States. The station will be capable of refueling each bus in less than 10 minutes. Total project costs are estimated to be $8.9 million. In addition to the FTA grant, other funding is being provided by Washington Natural Gas Company and the Washington State Energy Office. The 50 buses are scheduled to be on the road in 1993.

If it moves ahead on schedule, a project sponsored by the Sacramento Regional Transit District could be even larger than the Pierce Transit undertaking. Approved in March 1992, the project calls for the purchase of 75 natural gas buses at a cost of about $280,000 each. As with the Pierce Transit project, most of the cost of the Sacramento buses will be paid by FTA. Bus Industries of America will build the buses and begin delivery in March 1993 at a rate of five per week. In June 1992, a contract was signed for the construction of a $3.2 million natural gas refueling station in Sacramento.

Profile 7 Liquefied Natural Gas in Vehicles: Houston Metro and Roadway Express

Except for a small handful of vehicles, all of the 700,000 natural gas-burning cars, buses, and trucks in operation worldwide use compressed gas cylinders to store natural gas. A major drawback of current natural gas vehicle technology is that, by comparison to gasoline storage systems, conventional natural gas storage cylinders are heavy and bulky. (Profile 8 discusses these problems in more detail.) These problems are significantly reduced, however, by an alternative method of natural gas storage: use of liquefied natural gas (LNG). Once liquefied, natural gas can be stored at low pressure (10 to 25 pounds per square inch [psi], as compared to the 3600 psi standard for compressed gas cylinders) in insulated tanks that are only half the size and weight of compressed gas tanks holding the equivalent amount of fuel.

Although LNG storage is still bulkier and heavier than gasoline systems, it is a considerable improvement over compressed gas storage cylinders. A storage tank filled with LNG containing the energy equivalent of 20 gallons of gasoline weighs about 200 pounds, compared to about 500 pounds for a compressed gas storage system and 110 pounds for a standard car fuel tank filled with gasoline.

In addition to reducing space and weight, LNG vehicle storage systems offer other potential advantages over conventional gas storage cylinders. Refueling proceeds faster with LNG than with compressed natural gas. Moreover, liquefaction and shipment in LNG tankers offer the most economical method to transport natural gas across oceans. If, in the long-term, the United States wants to export its own natural gas or requires overseas imports of natural gas, LNG is the ideal form of the gas to use.

Challenges of Using Liquefied Natural Gas

Several technical, economic, and safety challenges must be met before use of LNG is widespread. Foremost among these challenges is the energy consumption required to reach and maintain the very low temperatures necessary to liquefy natural gas, and the associated costs. Natural gas liquefies at minus 258°F; to reach this temperature, it is necessary to use cryogenic technology. (Cryogenics, a branch of physics dealing with the behavior of materials at very low temperatures, is a stepchild of the space program: the liquid hydrogen and oxygen fuel systems used in space rockets have provided the basis for cryogenic research and technology development.) Applying cryogenics to natural gas liquefaction involves a combination of high-powered refrigeration and compression that can consume 25 percent of the energy content of the natural gas being liquefied. This adds approximately $0.20 per equivalent gallon of gasoline to the price of natural gas, thereby erasing much of the price advantage natural gas enjoys over gasoline.

In addition, once natural gas is liquefied, a major technical problem is that it is difficult to keep it cold. LNG is stored in cryogenic tanks which normally consist of a stainless steel inner tank and a steel or aluminum outer tank. The space between the tanks is filled with very efficient insulating material, such as alternating layers of aluminized mylar and silk netting. All air is evacuated from this space between the outer and inner tanks, thereby further insulating the LNG.

Nonetheless, even space age technology cannot prevent the gradual warming of LNG as it sits in a cryogenic storage tank. Eventually the LNG warms and begins to evaporate or "boil-off," and the pressure rises inside the tank. Boil-off in the best insulated cryogenic tanks is limited to about 1 percent of the liquefied fuel per day, and up to 5 percent in less efficient tanks. LNG storage tanks can handle about one week of accumulated boil-off before the pressure in the tank rises to unsafe levels that must be reduced.

If LNG-powered vehicles sit idle too long, the buildup of evaporated natural gas can present safety problems. As long as an LNG-powered vehicle is driven more than once per week, routine usage is sufficient to prevent a buildup of evaporated natural gas. The engine simply burns the evaporated natural gas before it begins to burn LNG. However, if such a vehicle sits idle for longer periods, the unburned fuel must be discharged to reduce pressure.

Technologies are available to protect against the potential LNG safety problem of evaporated natural gas buildup. For instance, storage systems can be fitted with pilot lights that periodically burn evaporated fuel in the tank, thereby reducing pressure. As of late-1992, however, the reliability and durability of such LNG safety features has not been assured.

A second safety concern stems from special precautions needed during the handling and storage of LNG at refueling stations. The super-cold liquid can, for example, cause frostbite if accidentally spilled onto a worker's hand during refueling. Moreover, as a liquid, LNG would disperse much more slowly than compressed natural gas in the event of a major leak or spill. Depending on the atmospheric conditions, this could increase the chance that LNG would ignite in the event of an accident. Safety standards are being developed to address these issues, however, and the overall safety risks of LNG are unlikely to prove to be more severe than for gasoline.

Houston Metro and Roadway Express

Liquefied natural gas-powered vehicle projects have existed since 1967, when the first LNG-powered vehicle was operated by San Diego Gas and Electric Company. A few other United States natural gas utilities have tested LNG-powered vehicles since then. In the late 1980s, the Canadian Energy, Mines and Resources Department sponsored a program to test the use of LNG in several heavy-duty trucks at a mining site in British Columbia. In Australia, a project begun in 1989 is currently testing an LNG-fueled heavy-duty truck for long-distance trips to the remote central regions of the continent. The LNG is obtained from a liquefaction plant in Alice Springs.

Two of the largest United States LNG projects are those involving Houston Metro and Roadway Express. In 1990, the Harris County (Texas) Metropolitan Transit Authority, commonly called Houston Metro, announced its commitment to convert more than 300 of its transit buses to LNG use, making it the country's largest alternatively fueled transit bus fleet. The decision to convert to LNG use followed completion of a demonstration project involving two 28-foot LNG-powered buses and eight compressed natural gas buses. The demonstration vehicles revealed size and weight advantages of LNG relative to compressed natural gas. As part of the Houston Metro project, the eight compressed natural gas buses used in the demonstration will be converted for LNG use.

The cab of a Roadway Express moving truck being refueled with liquefied natural gas. Photo: Roadway Express, Inc.

As of October 1992, Houston Metro's LNG fleet consisted of five buses. The fleet is projected to grow substantially in 1993 with additional bus deliveries under contracts calling for 36 articulated buses built by Neoplan (buses consisting of two units connected by accordion-like joints), 20 Mercedes buses, and 85 buses converted by the Stewart & Stevenson Company. In the longer term, another contract calls for the construction of another 180 LNG buses by Ikarus. Houston Metro estimates that the cost of each LNG bus is $35,000 more than the cost of a conventional diesel-powered model.

In 1990, Roadway Express, based in Akron, Ohio, began to study LNG for possible use in its truck fleet. The company operates 3500 long-haul moving vans and 5800 local moving trucks that annually consume 85 million gallons of diesel fuel. Roadway Express' study projected a test program to evaluate the performance of LNG in a 10-truck demonstration fleet. The first three trucks began operating on LNG in May 1991, and another four vehicles were converted to LNG in 1992. LNG is being supplied by a $250,000 refueling station built by Consolidated Natural Gas Company of Pittsburgh at Roadway's truck terminal in Copley, Ohio. The average cost of converting each truck to LNG is about $20,000, and the total cost of the project is estimated to be about $1 million.

Profile 8 Alternative Fuel Storage Technology: Institute of Gas Technology

Natural gas is slightly lighter than gasoline on an equivalent energy basis: that is, a pound of natural gas provides slightly more energy than one pound of gasoline. Thus, it is the weight of the fuel storage cylinders, not the fuel itself, that causes the "weight penalty" for natural gas vehicles. A conventional gasoline tank weighs just over 100 pounds, while an equivalent compressed natural gas cylinder storage system can weigh as much as 500 pounds. Moreover, the cylinders consume about four times the volume of a gasoline tank. The weight and volume of this storage system are needed to contain the compressed natural gas at high pressures (currently 3000 - 3600 pounds per square inch [psi], compared to atmospheric pressure of 14.7 psi).

The costs associated with conventional compressed natural gas storage cylinders are also considerable. Cylinder weight can cause a net reduction in fuel efficiency in natural gas vehicles of 7 percent compared to otherwise identical gasoline-powered vehicles. Additionally, in order to handle cylinder weight, the suspension systems in some vehicles may need upgrading. The cylinders themselves are an expensive item, accounting for up to half of the typical $2500 cost of converting a vehicle to natural gas use. Compressing natural gas to 3600 psi for storage accounts for about $0.20 of the consumer's cost for natural gas. Furthermore, cylinders require much of the trunk space in a converted natural gas vehicle.

To overcome the size and weight problems, most of the 30,000 natural gas vehicles on United States roads today have typically used one or both of two basic strategies: limiting the amount of natural gas storage or using both natural gas and another fuel in the same vehicle. Limiting the amount of natural gas fuel stored on board often means equipping a natural gas vehicle with only two natural gas storage cylinders, each capable of holding the equivalent of about five gallons of gasoline. These take up only about twice the space of conventional gasoline tanks and add less than 200 pounds to the vehicle (compared to a gasoline tank). However, such a natural gas vehicle's operating range is limited to less than 200 miles.

The alternative response to the size and weight problems is to build bi-fuel vehicles that retain the gasoline fuel tank and the ability to burn gasoline after conversion. Bi-fuel vehicles can revert to gasoline operation after the limited natural gas supply is depleted. Because the natural gas storage system is added to the gasoline storage system, rather than replacing it, bi-fuel vehicles equipped with two natural gas storage cylinders weigh around 300 pounds more than gasoline-fueled vehicles.

Using these strategies, natural gas conversions have proven most effective on vehicles that have sufficient room and heavy-duty suspension systems to handle natural gas cylinders, such as pickup trucks. However, improvements in natural gas vehicle fuel storage technology itself are needed if natural gas vehicle use is ever to become widespread.

During the past decade, natural gas fuel storage technology has advanced in two areas: increasing storage pressure (so that cylinders can accommodate more gas) and decreasing cylinder weight. A third advance, storing natural gas on molecules of carbon adsorbent materials, promises to offer an alternative to conventional compressed storage systems. However, adsorbent technology is currently in an early research stage and years away from commercial application in vehicles.

Until the mid-1980s, most United States natural gas cylinders were constructed of pressed steel and filled to pressures of 2400 psi, the standard for gas storage cylinders used in stationary-source energy applications such as on-site industrial gas storage. As natural gas vehicle interest has grown in the United States, however, new tanks have been designed that are capable of holding gas at 3000 and 3600 psi. Engineers, in turn, have developed refueling equipment capable of delivering natural gas to vehicles at these higher pressures. Thus, new tank pressure and refueling systems have increased average natural gas vehicle storage capacity by 25 percent without increasing cylinder weight, simply by increasing the natural gas content per tank. By mid-1992, 3000 psi had become the standard for most new natural gas vehicle projects, and tanks and refueling technology designed to compress and store natural gas at 3600 psi were beginning to enter the marketplace.

There have also been developments in connection with the commercialization of lighter-weight storage cylinders. Aluminum cylinders, which first appeared in large numbers in the mid-1980s, weigh about one-third less than pressed steel tanks that hold the same amount of natural gas. In the 1990s, composite-wrapped cylinders, which weigh as little as half as much as a conventional steel tank, entered the marketplace. These consist of light-weight aluminum or plastic cylinders which are wrapped tightly with very strong fiberglass or carbon filaments. The cylinder walls can be thin, and hence very light-weight, because they do not resist the pressure of the natural gas. They only provide the frame for the wrapping, which itself is strong enough to contain the compressed gas. Storage tanks are now available in a wide range of sizes and prices: in general, as the weights decrease from steel to aluminum to composite cylinders, the costs increase.

Carbon Adsorbent Technology

The third approach to natural gas fuel storage involves storing natural gas at lower pressures, with the aid of adsorbent carbon materials, rather than trying to compress more gas into lighter containers. Although it may seem counterintuitive, a cylinder that has been packed with certain types of activated carbon materials can hold a greater amount of natural gas at a lower pressure level than can a cylinder packed under high pressure with natural gas, but with no carbon adsorbents. The reason for this is that natural gas chemically reacts with the carbon, becoming adsorbed onto its surface. It takes up far less space when bound to a solid than in its gaseous state, even when the gas is highly compressed.

This van's natural gas cylinder storage system is packed with adsorbent carbon materials. Photo: Institute of Gas Technology

Adsorbent natural gas (ANG) technology is attractive for several reasons. First, operation of ANG systems does not require the expenditures associated with compressing gas to pressure levels of 3000 psi or more. Indeed, studies by the Chicago-based Institute of Gas Technology (IGT) indicate that ANG systems can reduce compression costs by half compared to conventional natural gas compression systems. Second, low-pressure ANG systems may eventually prove to be the storage technology most appropriate for home natural gas refueling systems (see Profile 13 for more information about home refueling). Third, because natural gas binds to solid carbon particles in ANG systems, the risk of a fire or explosion in case of a natural gas vehicle accident is reduced compared to compressed natural gas or liquid gasoline storage systems.

As of mid-1992, ANG systems that had been developed worsened the space problems associated with compressed natural gas use, making compressed gas storage a preferable alternative until ANG technology is further perfected. New materials, however, are being tested which may have the ability to hold even more natural gas per unit of space than is practicable with high-pressure gas storage cylinders.

The Institute of Gas Technology has been conducting the most extensive United States research program into ANG storage systems. Founded in 1941, IGT is a nonprofit energy research and education organization specializing in natural gas technologies. IGT has approximately 150 energy-related companies as full members and about 40 international associates. The bulk of IGT's annual budget of approximately $17 million (90 percent) comes from direct research grants and contracts with private and public sector clients; membership dues provide the remaining funding. The research is performed by IGT's 250 employees.

In 1982, IGT began carbon adsorbent research for use in natural gas vehicles; the research began as an offshoot of earlier hydrogen storage research for the National Aeronautics and Space Administration (NASA). Since then, IGT has tested dozens of carbon-based materials for their capacity to bind with, store, and eventually discharge natural gas. The goal of the current program, which is partially funded by the Gas Research Institute, is to identify carbon-based materials that can hold, by adsorption in a carbon-packed cylinder at 500 psi, about 60 percent of the natural gas contained in a conventional cylinder at 3000 psi. IGT is also conducting research projects to design and test different ANG storage vessels.

Research into ANG technologies must also develop solutions to difficulties encountered during refueling. It takes time for natural gas to permeate activated carbon that is densely packed in a cylinder. Moreover, heat is released as natural gas is adsorbed during refueling, and this heat must be continuously removed during the refueling. Another obstacle is posed by deterioration or contamination of the surface of the activated carbon. Deterioration or contamination reduces the fuel-binding effectiveness, or cyclability, of activated carbon over time. IGT hopes to overcome all these technological barriers. It is, for example, testing a variety of heat exchange systems capable of rapidly cooling the activated carbon so that rapid adsorption can occur.

The Institute of Gas Technology is not the only organization researching adsorbent technologies. ANG research is also underway at Syracuse University (New York) and at a Union Carbide Corporation laboratory in New York. Additionally, the Atlanta Gas Light Company is a project coordinator for an effort to develop and test an ANG system in four Chevrolet S-10 pickup trucks.

Profile 9 Natural Gas and Hydrogen Fuel Mixtures: The Denver Hythane Project

Despite the considerable environmental, safety, energy security, supply, and cost benefits of natural gas, its use raises at least two serious concerns. First, as with oil-derived fuels, the supply of natural gas is limited and nonrenewable. Second, widespread use of a gaseous fuel, as opposed to a liquid fuel, would require substantial changes in the nation's fuel transportation infrastructure.

However, these concerns about natural gas use might not prove to be long-term liabilities if a transition to natural gas vehicles eases the transition to the virtually inexhaustible and clean-burning hydrogen fuel produced by using solar energy to break water molecules into their hydrogen and oxygen components. Hydrogen combustion is virtually pollution-free because its primary by-product is pure water; thus, burning hydrogen is even more environmentally desirable than burning pure natural gas.

Viewed from this perspective, natural gas vehicles may be vital to the successful transition to hydrogen fuel. Because natural gas vehicle technology is more advanced than hydrogen fuel technology, use of natural gas vehicles will help establish the infrastructure and technology necessary for ultimate utilization of hydrogen. The unlimited availability of solar-derived hydrogen, in turn, provides a long-term policy rationale that helps justify present natural gas vehicle investment.

The transition from natural gas to hydrogen is a logical one in several respects. First, for example, the southwestern United States is both the major natural gas-producing region in the country and the region that receives the most sunshine. Therefore, natural gas fields, once abandoned, are possible sites for solar-powered hydrogen production facilities. Second, the existing natural gas pipeline system, more than 1.1 million miles long, which emanates largely from the Southwest, can eventually and easily be converted to carry hydrogen gas to markets. Third, advances in natural gas storage technology for natural gas vehicles are likely to provide applications for hydrogen fuel storage in vehicles.

Until dedicated hydrogen vehicles are more widely available and technically and economically feasible, an alternative exists to the use of pure hydrogen. A natural gas-based fuel mixture containing up to 20 percent hydrogen can be burned in an engine designed for burning pure natural gas without significantly changing engine behavior, damaging the engine, or hurting engine performance. Use of natural gas and hydrogen fuel mixtures provides an opportunity both to introduce hydrogen as a transportation fuel and to demonstrate that natural gas use can serve as an effective transition to a renewable, hydrogen-based energy future.

The Denver Hythane Project

A project now underway in Denver is exploring the potential use of mixtures of natural gas and hydrogen fuels in vehicles. The Denver project represents the first major United States demonstration project testing the operation of vehicles fueled by these fuel mixtures in routine driving conditions.

A hythane-fueled vehicle operates at Stapleton International Airport as part of the Denver Hythane Project in Colorado. Photo: Denver Department of Health & Hospitals

In 1987, facing an alarming increase in air pollution in Denver and other cities along the Front Range of the Rocky Mountains, the state of Colorado initiated an aggressive alternative transportation fuels program. The program focuses mainly on the use of gasolines enriched with oxygenated fuel additives and the expanded use of natural gas and methanol vehicles. In 1990, however, the program was expanded to include a demonstration project designed to evaluate the performance of vehicles fueled by a mixture of 15 percent hydrogen and 85 percent natural gas. The mixed fuel is referred to as hythane.

Denver's hythane project is managed by the Denver Department of Health and Hospitals and funded by the US Department of Energy's Urban Consortium Energy Task Force, state and local government agencies, and a number of private sector partners. The project utilizes technology developed and installed by Hydrogen Consultants, Inc., based in Littleton, Colorado, one of the few companies in the United States engaged in the development of hydrogen vehicle technology.

Vehicles receive a mixture of natural gas and hydrogen at this Denver Hythane Project refueling station. Photo: Denver Department of Health & Hospitals

The Denver Hythane Project is being conducted in three phases. Phase I, which has been completed, involved the testing of one Chevrolet S-10 pickup truck converted to run on hythane. The results indicated normal vehicle performance and impressive reductions in exhaust gas pollution, as measured at three certified testing facilities in Colorado and California. Specifically, hythane use reduced hydrocarbon emissions by 50 percent and nitrogen oxide emissions by 66 percent when compared to the operation of the vehicle on pure natural gas. This is a significant improvement over the already impressive pollution control advantages of natural gas as compared to gasoline (see Chapter 1). The Denver hythane test therefore indicates that addition of even small quantities of hydrogen to natural gas can produce large additional reductions in air pollution.

Phase II of the study, which began in the summer of 1991, involves a comparison of the performance of three pickup trucks, including a gasoline-powered truck operated by the Denver Water Board, a natural gas-powered truck operated by Public Service Company of Colorado, and a hythane-fueled vehicle operated as part of the Stapleton Airport vehicle fleet. The demonstration program calls for air pollution emission measurements to be taken for each truck after 4000, 7500, 10,000, and 25,000 miles of driving.

Phase III of the hythane project, beginning in 1993, will involve comparative testing of natural gas and hythane fuels in three dedicated light-duty trucks converted for natural gas use.

In May 1992, the Denver Hythane Project was one of the winners of the Pollution Prevention Awards made by the Administrator of the US Environmental Protection Agency. Prompted by the success of the Denver project, hythane programs are being developed by the State Energy Office in Pennsylvania and by the Florida Solar Energy Center.

Chapter 4 Natural Gas Vehicle Commercialization and Refueling Infrastructure Development

Chapter 3 outlined the kinds of technical research, development, and demonstration programs that are gradually providing evidence of the environmental, economic, and safety advantages of natural gas vehicles. Once natural gas vehicles are thus shown to be viable from these vantage points, the next challenge is creating market demand sufficient to bring about widespread commercial natural gas vehicle production.

Market development will require considerable effort and investment. Because of the comparatively small number of natural gas vehicles and natural gas vehicle refueling stations, widespread commercialization and the creation of an appropriate refueling infrastructure are frustrated by the proverbial "chicken and egg" syndrome.

On the one hand, automotive manufacturers are reluctant to invest in commercial-scale natural gas vehicle production due to low consumer demand, which is in turn restrained by the fact that it is difficult, if not impossible, for most private individuals to refuel a natural gas vehicle. On the other hand, investments in natural gas refueling stations are held back by the lack of natural gas vehicle drivers to patronize them. Retail automotive refueling is traditionally a business with a low profit margin: large quantities of fuel must generally be sold at each location to generate the income needed to pay the cost of the refueling equipment. Retail fuel sellers, therefore, are reluctant to invest in refueling facilities until a large population of natural gas vehicles is on the road.

In contrast to the nascent natural gas vehicle market (with only 30,000 natural gas vehicles on United States roads), the market for gasoline-powered vehicles is efficient and established. On average, 11 million vehicles are sold annually in the United States. However, these numbers are attainable only because of the economies possible with assembly line automobile manufacturing. For example, profitable car manufacture generally requires annual production runs of 100,000 vehicles for each car model.

Moreover, development of new vehicle models is an expensive proposition. After completion of basic research and development, the design and equipment retooling needed to manufacture 100,000 units of a new product line typically requires a five-year commitment that can easily cost more than $1 billion. In short, the financial costs of car manufacture are so high that the incentive to pursue new technologies must be great.

Commercial development of natural gas vehicles is also hindered by a refueling infrastructure that only accommodates the ever-increasing numbers of gasoline and diesel vehicles on United States roads. Because automobiles can travel only relatively short distances between refuelings, this infrastructure is massive. Conventional vehicles can carry only enough fuel on board to drive about 300 miles. Since United States motorists drive more than two trillion miles each year, this translates to at least 6.67 billion visits to gas stations.

Each gas station visit entails use of equipment that is part of the refueling infrastructure put into place over decades — at a cost of tens of billions of dollars. To service the tremendous thirst for fuel, there are more than 100,000 gasoline stations and privately owned fuel pumps. By contrast, there are only about 500 natural gas refueling stations, and virtually all of these are privately owned and used solely to refuel commercial fleets of natural gas vehicles. Without more vehicles to patronize refueling stations, retail sales of natural gas fuel are unlikely to expand.

Overcoming the obstacles of a lack of market demand for natural gas vehicles and the absence of a refueling infrastructure is vital if the widespread benefits of natural gas vehicles are to be realized. This chapter discusses five of the key actions needed to do so. The first two stimulate commercial demand and promote product reliability for natural gas vehicles. The remaining three help establish a refueling infrastructure to service natural gas vehicles.

Natural Gas Vehicle Commercialization and Refueling Infrastructure Development

What Needs To Be Done	Key Actors
10. Developing large purchase orders for natural gas vehicles	Ten-utility natural gas vehicle purchase consortium and federal efforts
11. Assuring reliable aftermarket conversion kits	American Gas Association Laboratories
12. Establishing a refueling station infrastructure	Natural Fuels Corporation and the Natural Gas Vehicle Zone
13. Demonstrating vehicle refueling appliances	FuelMaker Corporation
14. Removing restrictions against retail natural gas sales	Federal and state governments

Profile 10 Developing Large Purchase Orders for Natural Gas Vehicles: Ten-Utility Natural Gas Vehicle Purchase Consortium and Federal Efforts

Alternative-fuel research programs typically involve only a few test vehicles, and demonstration programs commonly showcase only a limited number of vehicles. Full-scale commercialization of natural gas vehicles will only be achieved when thousands of natural gas vehicles are available for sale in large United States automotive sales markets.

One way to build a large market is by encouraging large purchase orders for natural gas vehicles. Large purchase orders during the next few years will support the continued economic viability of manufacturing conversion kits (parts needed for converting standard gasoline-powered vehicles to vehicles that can burn natural gas). They will also encourage automotive manufacturers to retool for assembly line production of alternative fuel vehicles, thereby promoting economies of scale and lowering the cost of each vehicle.

Ten-Utility Natural Gas Vehicle Purchase Consortium

As of mid-1992, the most successful program to build commercial market demand for natural gas vehicles had been undertaken by a consortium of ten natural gas utility companies. In addition, a recently expanded federal government program could further bolster the growing natural gas vehicle commercial market.

During the first half of 1991, several natural gas utility companies in the western United States developed a joint program to purchase natural gas vehicles produced by a single manufacturer. Ten companies, located in California, Colorado, and Texas, ultimately formed a consortium for this purpose. They include:

California
- Pacific Gas & Electric Company (San Francisco)
- San Diego Gas and Electric Company (San Diego)
- Southern California Gas Company (Los Angeles)

Colorado
- Natural Fuels Corporation (Denver)

Texas
- El Paso Natural Gas Company (El Paso)
- Energas Company (Midland)
- Enron (Houston)
- Entex (Houston)
- Lone Star Gas Company (Dallas)
- Southern Union Gas Company (Dallas)

In mid-1991, the consortium negotiated a purchase order with General Motors Corporation: the consortium paid GM $935,000 toward the cost of engineering and manufacturing a dedicated natural gas-burning pickup truck and agreed to purchase at least 1000 of these natural gas vehicles in the 1991 model year. For its part, GM agreed to build the vehicles and to fully warranty the pickups as if they were conventional gasoline-burning vehicles. GM also agreed to distribute the natural gas vehicles through its network of dealerships, thereby incorporating its extensive automotive distribution and sales network into the natural gas vehicle business for the first time. As discussed in Profile 2, the Sierra pickup truck equipped with a 5.7-liter engine was selected as the production model under the consortium's program and has been certified as meeting California's strict 1992 emission controls.

Publicity surrounding the announcement of the consortium-sponsored purchase order generated considerable interest. Within a few months, GM had received orders for an additional 1000 natural gas pickups. Late in 1991, about the time that customers began receiving the first natural gas vehicles, GM announced plans to market up to 10,000 natural gas-burning Sierra pickups by the end of the 1992 model year. Thus, this one 1000-vehicle purchase order could stimulate additional sales that will increase the total number of natural gas vehicles on United States roads by over 30 percent.

Federal Efforts

Until 1992, the federal government had done little to encourage the use of alternative transportation fuels through its own vehicle purchasing decisions. A few programs had been implemented by the US Department of Energy (DOE) under the provisions of the Alternative Motor Fuels Act of 1988 which authorized a $16 million alternative transportation fuels program.

In 1991, DOE drew fire from several quarters, including the US General Accounting Office and the Energy Subcommittee of the House Government Operations Committee, for not acting more aggressively. Congressional hearings revealed that DOE had converted only 115 vehicles under the three-year-old federal law. Purchases of alternatively fueled vehicles by other federal agencies were similarly meager.

On April 19, 1991, President Bush issued Executive Order #12759 which, coupled with the requirements of the Energy Policy Act of 1992, will significantly expand federal purchases of vehicles powered by alternative transportation fuels. Section 11 of the Executive Order requires that the "maximum number practicable of vehicles acquired annually (by federal government agencies) are alternative fuel vehicles." In response to the order, the federal government's chief purchasing agent (the General Services Administration, or GSA) and other agencies began ordering more alternative fuel vehicles. By early 1992, the GSA had announced purchase orders for 3125 alternatively fueled vehicles, including 600 dedicated natural gas vehicles, in 1992.

As of mid-1992, the largest federal natural gas vehicle fleet was operated by the US Postal Service. By the end of 1991, 314 vehicles, mostly half-ton delivery vans, had been converted to run on natural gas. By the end of 1992, however, an additional 1000 vehicles will be converted to natural gas use.

As the GSA increased its purchases of alternatively fueled vehicles in 1992, the Department of Energy developed a five-year vehicle purchase plan to fully implement the executive order. The work product was a report released in August 1992 entitled *Alternative Fuel Vehicles for the Federal Fleet: Results of the 5-Year Planning Process.* The report proposes a vehicle procurement strategy to ensure that the following targets are met: 5000 alternative fuel vehicle purchases in 1993, 7500 purchases in 1994, and 10,000 in 1995. By 1998, 50 percent of federal fleet purchases would be alternatively fueled. Although developed in response to the President's executive order, the vehicle acquisition schedule in the DOE five-year plan became a legal requirement when it was incorporated by Congress as a federal government mandate through a provision of the Energy Policy Act enacted October 24, 1992.

Postal Service employees refuel a delivery truck with natural gas. Photo: US Postal Service

Specifically, the five-year plan calls for the acquisition of 5707 alternative fuel vehicles in the federal 1993 fiscal year. Of these, 2652, or 46 percent, will be natural gas vehicles. Most of the rest, some 2865 vehicles, will be alcohol-fueled, either by methanol or ethanol. By 1997, the number of natural gas vehicles purchased by the federal government is projected to grow to nearly 5000.

The federal government is also attempting to increase purchase orders for alternative fuel vehicles by linking federal and state government vehicle procurement programs. In June 1992, DOE Assistant Secretary Michael Davis requested information from the energy offices in all 50 states about state government programs to purchase alternative fuel vehicles. It is the intention of the DOE program to combine purchase orders for similar types of vehicles from federal and state agencies and to present these as single vehicle orders to automotive manufacturers.

The US Postal Service operates the largest federal natural gas fleet. Photo: US Postal Service

Profile 11 Assuring Reliable Aftermarket Conversion Kits: American Gas Association Laboratories

In the long term, a future in which natural gas vehicles are widely used will require mass production of cars and trucks designed specifically to maximize the advantages of natural gas as a transportation fuel. In the shorter term, however, wider availability of natural gas vehicles depends upon an "aftermarket" that converts vehicles manufactured to burn gasoline or diesel fuel into vehicles capable of burning natural gas. The aftermarket conversion industry is composed of companies that build conversion kits and install them in conventional vehicles.

The 1992 edition of the *Directory of Natural Gas Vehicle Refueling Stations, Products & Services*, published by the American Gas Association (AGA), reports that there are approximately 180 AGA-registered companies in the United States selling parts or performing vehicle conversions to natural gas use. This is more than twice the number of companies listed in the 1991 directory, attesting to the rapid growth of the natural gas vehicle industry.

The basic engineering principles on which conversion systems are based are the same, but each conversion kit on the market is unique. Although all conversion kits contain storage tanks, pressure regulators, fuel mixers, mountings and brackets, and various other minor parts, the parts and design of conversion kits are not standardized, and the number of product differences inevitably results in a wide range of performance characteristics. As a result, some conversion systems work better than others and some — improperly installed or containing undetected production flaws — may not work well or for long at all. Further, many purchasers of natural gas conversion kits are so inexperienced in natural gas vehicle technology that they are unable to differentiate among the many conversion products offered for sale and are hard-pressed to make informed purchase decisions.

The availability of conversion kits of varying quality presents a problem for widespread commercialization of natural gas vehicles. It is therefore essential to provide buyers of conversion kits with some form of quality guarantee. To fill this need, natural gas conversion kit certification programs are being established and educational institutions are opening their doors to provide nuts and bolts natural gas vehicle training for mechanics and natural gas vehicle purchasers.

American Gas Association Laboratories Certification Program

The major nationwide certification program in effect is an effort spearheaded by the American Gas Association Laboratories in Cincinnati, with the assistance of the Natural Gas Vehicle Coalition. AGA's certification program began in 1984 with a standards development effort. In 1985, certification procedures were outlined in *A.G.A. Requirements for Natural Gas Vehicle (CNG) Conversion Kits*. However, companies working in the field of natural gas vehicle technology in these early years generally did not participate in this certification program. The small market for natural gas vehicle equipment sales did not justify the time and expense of complying with thc AGA certification procedure.

In 1989, the conversion kit certification idea gained new momentum with the completion of the first comprehensive review by AGA of codes and standards pertaining to natural gas vehicles and refueling stations. Information collected in that report provided the basis for new certification procedures. As a result, the 1985 program was updated and expanded to reflect the current state of natural gas vehicle technology.

AGA's certification process involves three basic components: reviews of conversion kit (1) construction, (2) performance, and (3) manufacturing, including site inspection. The construction component of the certification process evaluates the physical attributes of conversion kit products and parts, including strength, durability, and the adequacy of product labeling and instructional materials that accompany a conversion kit.

The performance component of the certification process tests conversion kit products against minimum operating criteria to ensure that they work properly. Tubes and fittings, for example, are subjected to four times their normal operating pressure of 3000 pounds per square inch in order to assess their ability to contain fuel safely without leakage or structural failure. Similarly, valves and regulators must be able to function without leakage or failure at pressures 1.5 times greater than the normal operating pressure and over a temperature range from – 40 to 250° F.

Finally, the manufacturing component of the certification process requires applicants to submit a "Manufacturing and Production Test Plan" explaining how their conversion kits are built, assuring that adequate in-house quality control procedures are followed, and identifying suppliers of materials and parts not produced directly by the conversion kit manufacturer. The manufacturing component also includes a site inspection of production facilities where kits are built, with quarterly inspections of certified manufacturers to follow thereafter. AGA Laboratories retains records from the certification process for seven years.

By October 1992, five conversion kit manufacturers had successfully completed the certification process: Automotive Natural Gas Inc. (Milton, Wisconsin), National Energy Service Company (Zanesville, Ohio), CleanFuels Inc. (Martinsburg, West Virginia), Metropane Inc. (Columbus, Ohio), and Propane Equipment Company. Conversion kits from these companies now display an AGA certification symbol.

Manufacturers of component parts used in a variety of conversion kits assembled by other companies have also applied for separate certification of their parts. To date, regulators, valves, and fuel mixers produced by IMPCO Technologies Inc., have been certified, as have storage cylinder brackets built by CNG Cylinder Company and Shoreline Compression Company and electronic equipment produced by Energy Kinetics Inc. and Hercules Engines Inc. Refueling nozzles built by Sherex Industries and the FuelMaker vehicle refueling appliance have also been certified.

State Certifications

Following the lead of AGA Laboratories, several states are developing their own certification standards and requiring certification of natural gas vehicle technology sold within their boundaries. California and Colorado have both established certification procedures, basing certification upon compliance with air pollution emission standards set by the California Air Resources Board (CARB). These states' certification requirements apply to three vehicle categories: automobiles, light-duty trucks, and heavy-duty trucks. The California and Colorado programs are directed by CARB and the Colorado Department of Health, respectively.

Texas has established a natural gas vehicle certification requirement based largely on compliance with product standards set by the National Fire Protection Association (NFPA), and a 1990 Oklahoma law requires natural gas vehicle technology in that state to meet certification standards similarly based on the NFPA Code 52. Code 52 of the NFPA sets product and design standards for compressed gas systems. As discussed in Profile 20, until regulatory standards are established specifically for natural gas vehicle technology, compliance with the NFPA Code 52 provides reasonable assurance that natural gas vehicle technology will operate properly and safely.

Training and Education

The need to train competent mechanics to install conversion kits and service and repair natural gas vehicles is related to the need to ensure and guarantee the product quality of natural gas vehicle conversion kits. In fact, many conversion kit manufacturers offer to train mechanics employed by purchasers. Training may address issues such as proper installation procedures, so that mechanics can convert additional vehicles in the future without the assistance of the conversion kit manufacturer. Alternatively, training may be limited to proper vehicle repair and maintenance procedures for natural gas vehicles.

Several institutions, including the Vehicular Fuels Institute located at the Hocking Technical College in Nelsonville, Ohio, now offer specialized vocational education programs focused on natural gas vehicle technology. These organizations offer multiday training courses addressing many aspects of natural gas vehicle conversion, maintenance, and use. The Community College of Santa Fe is one of the first colleges in the country to offer a credit course in natural gas vehicle technology. Development of the curriculum for this course was funded by the New Mexico Energy, Minerals and Natural Resources Department. Similar programs are being established in the state vocational training center in Oklahoma.

Profile 12 Establishing a Refueling Station Infrastructure: Natural Fuels Corporation and the Natural Gas Vehicle Zone

The lack of a natural gas refueling station infrastructure has presented a significant obstacle to the widespread commercialization of natural gas vehicles. However, creation of such an infrastructure is now underway. In recent years, natural gas vehicle refueling technology for compressed natural gas has been greatly improved. No major technological problems remain, although some fine tuning is needed and increased standardization of parts is greatly desirable. The major problems to be solved before an integrated refueling network is firmly in place are a reduction in both the high cost of refueling equipment and the long lead time required to establish such a network.

Most natural gas refueling systems employ dispenser pumps similar to those commonly used at retail gasoline stations. One difference is that a hose carrying natural gas is much thinner than a gasoline hose, and the nozzle linking a natural gas dispenser to an automobile fuel intake port makes a leakproof connection (much as a compressed air pump nozzle grips a car tire tube valve).

There are two basic types of natural gas refueling systems: fast-fill and slow-fill. Fast-fill systems use powerful compressors to fill large natural gas storage cylinders to 3600 pounds per square inch (psi) or more, a significant increase in pressure over the 3000 psi used in vehicle storage tanks. The fuel is drawn from existing natural gas pipelines, and storage cylinders, called a cascade, are grouped in a shielded area of the refueling station. When a refueling hose is connected from a storage cylinder to a vehicle, a release valve is opened and natural gas floods very quickly down the pressure gradient from the on-ground storage tanks to the vehicle. Refueling stops when the vehicle's on-board cylinder pressure reaches the desired pressure, usually 3000 psi. As refueling occurs, the compressors are activated to refill the on-ground storage cylinders.

During refueling, the flow of natural gas is measured as it passes through a fuel dispenser. A natural gas fuel dispenser can be designed to be virtually indistinguishable from a conventional gasoline fuel pump, complete with tumbling dials that measure accumulated fuel delivery and cost. The principal advantage of fast-fill refueling stations is speed: they can replenish natural gas vehicle storage cylinders in three to five minutes, only slightly longer than it takes to refuel the average car with gasoline.

The major drawback of fast-fill refueling is its comparatively high cost compared to slow-fill refueling, mainly resulting from the capital expense of the compressors and the storage cylinders: a retail fast-fill natural gas vehicle refueling outlet capable of serving 300 autos per day can cost over $400,000 (compared to less than $100,000 for a gasoline station). Larger and more powerful compressors needed to refuel fleets of transit buses can push the cost of a refueling station for these vehicles to over $1 million. In addition, considerable energy, usually electricity, is required to run the compressors. Between one and two kilowatt-hours of electricity are consumed during the compression of one equivalent gallon of natural gas (that is, the amount of natural gas containing the same amount of energy as one gallon of gasoline), at a cost of about $0.10. The capital costs add another $0.10 per equivalent gallon, cutting deeply into the price advantage uncompressed natural gas enjoys over gasoline. Nonetheless, even with the refueling capital and operating costs factored in, compressed natural gas still usually sells at $0.30 to $0.40 per equivalent gallon below gasoline cost, and refueling station operators can recoup their investment through sales over the course of several years.

By comparison to fast-fill refueling, the capital and operating costs of slow-fill refueling, which uses much smaller compressors to pump natural gas directly from the distribution pipeline into a natural gas vehicle's storage tank over a six- to twelve-hour period, are less than half those associated with fast-fill. The obvious drawback, however, is that natural gas vehicles must be immobilized, usually overnight, in order to be refueled. The application of slow-fill technology is limited largely to commercial fleet applications in which vehicles are routinely parked in a centralized location with adequate room at each parking slot to accommodate a slow-fill refueling apparatus.

Emerging Refueling Infrastructure

In 1989, there were fewer than 200 natural gas refueling stations in the United States. Virtually all of these were privately owned and operated by natural gas utility companies for use by their natural gas vehicle fleets and by private commercial natural gas vehicle fleets. Between 1990 and mid-1992, however, this number more than doubled. In 1991 alone, 156 refueling stations opened for business, an average of three per week. As of mid-1992, 511 natural gas refueling stations were listed in the American Gas Association *Directory of Natural Gas Vehicle Refueling Stations, Products & Services*. These refueling stations are located in 43 states and the District of Columbia. Colorado leads the way with the most refueling stations: 41. Ohio and California each have more than thirty stations: 34 and 32, respectively.

Although most refueling stations continue to be owned by natural gas utilities for their private use, utilities are increasingly opening their refueling stations for use on a pay-as-you-go basis by other natural gas vehicle users. For instance, on October 17, 1990, the nation's largest utility-owned refueling station opened to the public. Owned by Minneapolis-based Minnegasco, Inc., the station can simultaneously refuel four vehicles in less than five minutes. A credit card system is used to provide access to the fuel dispensers and record fuel consumption for billing purposes. Although the station cost about $500,000, Minnegasco expected to recoup its investment through natural gas sales at $0.60 per gallon equivalent in less than two years. The Minnegasco station currently services approximately 70 natural gas vehicle customers.

Another development involves the construction of natural gas refueling stations owned and operated by companies other than natural gas utilities, including major oil companies. For example, in 1990, Amoco became the first major oil company to provide natural gas refueling when it added natural gas dispensers to some of its retail gasoline stations in Colorado. Other oil companies, including Chevron, Phillips 66, Shell, Texaco, and Unocal, have since followed Amoco's example.

Natural Fuels Corporation

Natural Fuels is conducting a major effort with its partner, Vickers, to bring natural gas pumps to retail gasoline stations throughout Colorado. Photo: Natural Fuels Corporation

One of the most ambitious private sector natural gas refueling initiatives is underway at Natural Fuels Corporation, headquartered in Denver, Colorado. The company is a joint venture of the Public Service Company of Colorado, Colorado Interstate Gas Company, and Julander Platt Nelson Inc. (a private investing company).

Advertised as the nation's first "full-service alternative transportation motor fuel company," Natural Fuels opened its first public refueling station at Denver's Stapleton Airport on June 4, 1990. Its initial business plan aims to establish 80 refueling stations along Colorado's Front Range by 1996. Natural Fuels owned or operated about 20 refueling stations by October 1992. In addition to selling natural gas at its own stations under the "Natural" trademark, Natural Fuels has built two refueling stations as part of a joint agreement with Amoco. In 1992, Natural Fuels began a second collaborative effort with an oil company. This program involves installation by Natural Fuels of natural gas refueling equipment at 12 refueling stations currently operated by Vickers, a subsidiary of TOTAL Petroleum. As part of the promotion of this partnership, fleet operators who converted their vehicles to natural gas before the end of 1992 received $200 worth of free natural gas from the participating Vickers stations. Finally, Natural Fuels services about ten private refueling stations on behalf of the Public Service Company of Colorado.

To further encourage creation of a market of natural gas vehicles requiring refueling, Natural Fuels has also entered the vehicle conversion business, having opened its first conversion center in Denver in October

1990. During its first year of operation, 360 vehicle conversions were completed at the center. (See Profile 11 on efforts to provide reliable aftermarket vehicle conversion.)

Natural Fuels' business plan identifies Texas and California as the primary market areas into which the company will expand in the early 1990s, after establishing itself in Colorado. Secondary target markets include other southwestern states, Illinois, and Minnesota.

Southwestern United States Natural Gas Vehicle Zone

Recognizing the emergence of interest in natural gas sales by gas utilities and gasoline retailers, Texas Governor Ann Richards wrote to the governors of all southwestern states in January 1992 to seek cooperation in establishing a "Natural Gas Vehicle Zone." The objective of Phase I of this program is to create a network of retail natural gas refueling stations at 200-mile intervals along interstate highways from Louisiana to California. In Phase II, natural gas dispensers would be added to 50 percent of all automotive service stations in the Natural Gas Vehicle Zone.

By October 1992, seven southwestern states had joined the Natural Gas Vehicle Zone, and a steering committee had been established to develop a strategic plan for Phase I of the program. In the future, project participants hope to expand the zone to other parts of the nation and to include the three Mexican states that border the United States.

Profile 13 Demonstrating Vehicle Refueling Appliances: FuelMaker Corporation

Drivers of conventional, gasoline-powered automobiles and trucks are prisoners to refueling stations. Their everyday life is affected by the necessity of checking the fuel gauge needle and locating and patronizing a refueling station.

By contrast, the possibility of natural gas vehicle home or business refueling could revolutionize the refueling experience. Natural gas is the only alternative fuel with the potential to offer consumers a significant degree of freedom from the corner service station. Vehicle refueling appliances — a technology now on the market — allow individual consumers to refuel natural gas vehicles overnight, while a vehicle sits in a driveway or garage at a home or business.

With little effort, users of vehicle refueling appliances can leave their homes each morning with a full tank of natural gas or arrive at work to find a business car with the fuel gauge reading full. Billing is automatic, as fuel charges are merely added to the monthly natural gas statement a consumer receives from the local utility. An added advantage of this refueling system is that it reduces energy consumption and traffic congestion by eliminating trips to purchase fuel. Moreover, if home or business refueling becomes widespread, some of the thousands of roadside properties dedicated to conventional service stations could be recovered, permitting a better-planned approach to land use and leading to a more attractive landscape. While refueling stations would be required to allow long-distance travelers, drivers away from home, and apartment dwellers to refuel their vehicles, fewer would be necessary.

Small-scale vehicle refueling appliances use natural gas tapped from a home or business hook-up. They are an offshoot of slow-fill refueling technology. Natural gas is compressed slowly and fed directly into the vehicle. While slow-fill systems can be designed to simultaneously refuel entire fleets of vehicles overnight, home refueling appliances are small, self-contained units designed to refuel only one or two vehicles. Thus, vehicle refueling appliances are most appropriate for home or business owners who operate only one or two vehicles.

There are at present over 1.1 million miles of natural gas pipelines in the United States, delivering fuel across the nation to locations where about 95 percent of the population lives and works. This population is a ready market for natural gas vehicle refueling appliances. Indeed, the creation of an individual refueling network serving even a small segment of this market could have a revolutionary impact on the country's refueling patterns.

The FuelMaker

The company pioneering vehicle refueling appliance technology is Sulzer Brothers Ltd., headquartered in Winterthur, Switzerland. A world leader in the production of industrial compressors and conventional natural gas refueling systems, Sulzer undertook a five-year research and development program in the early 1980s to perfect a small-scale vehicle refueling appliance.

By the mid-1980s, Sulzer's effort had resulted in the manufacture of a prototype refueling appliance. In 1987, Sulzer began test marketing a prototype home refueling appliance. More than 250 units were distributed for demonstration use in homes in Europe (100 units), Canada (90), Australia (60), and New Zealand (30). Although these units generally performed adequately, they experienced mechanical problems, were quite noisy, and were somewhat cumbersome to use. Experience gained during the testing of these units, however, assisted Sulzer in the design of later models with much improved performance characteristics.

As Sulzer's undertaking grew, the vehicle refueling appliance technology was transferred to the FuelMaker Corporation, based in Vancouver, Canada. The Fuelmaker Corporation is a partnership established in 1989 by Sulzer, BC Gas Company of Vancouver, and Questar Corporation of Salt Lake City. Since then, it has handled the development, production, and marketing of the vehicle refueling appliance, called the FuelMaker.

In 1990, FuelMaker Corporation began to market its most advanced home refueling model, the FuelMaker C3. In December 1990, the FuelMaker C3 received a "Certification Seal for Appliances" from the American Gas Association Laboratories, in recognition of its safety and reliability. This certification opened the door to marketing in the United States.

Natural gas vehicle owners can use the FuelMaker refueling appliance at home. Photo: FuelMaker Corporation

The basic mechanism of the FuelMaker is a small compressor, not much bigger than the unit commonly used to power large home refrigerators. There is no compressed gas storage capacity within the FuelMaker. Natural gas from the conventional residential hook-up is piped into the FuelMaker, where it is gradually compressed to 3000 pounds per square inch (psi); a refueling nozzle links the FuelMaker to the natural gas vehicle refueling port, and the compressed gas is pumped directly into the vehicle. In its entirety, the FuelMaker is only 28 inches high and long, and weighs just 145 pounds. A simple white metal cover hides the mechanical apparatus and helps to reduce compressor noise.

The FuelMaker can compress about one equivalent gallon of natural gas per hour into a natural gas vehicle. At the end of a day, a driver need only pull the vehicle abreast of the FuelMaker, attach the refueling nozzle, and turn on the compressor switch. The machine turns off automatically when the vehicle's storage tanks are filled. By refueling their cars overnight, some natural gas vehicle owners can thus benefit from the less expensive, off-peak electricity rates offered by some utility companies. In sum, as the FuelMaker is currently designed, it is most suitable for vehicles driven less than 150 miles a day and housed overnight at a residence or business already served by natural gas.

FuelMaker's marketing strategy in the United States has been to sell demonstration vehicle refueling appliances to natural gas utilities. The utilities in turn sell or lease the units to hand-picked customers, usually businesses or government leaders, for use in demonstration programs. In August 1991 the company manufactured the 1000th C3 series FuelMaker. By mid-1992, over 2000 units had been delivered worldwide or were on order.

Building on the success of the C3 technology, FuelMaker began commercial production of a new generation of its vehicle refueling appliance with the unveiling of its FM4 model in the spring of 1992. The FM4 reflects a number of design modifications and engineering advancements. Most importantly for the consumer, the interval between required routine servicings has been extended from 1000 hours of operation for the C3 to 2000 hours for the FM4.

Vehicle refueling appliances currently have several drawbacks. One is cost. The FuelMaker C3, for instance, is priced at around $3500, and installation costs another $500. Another concern with vehicle refueling appliances is durability. The mechanical performance of vehicle refueling appliances has not yet been perfected. Although there have been no serious safety problems associated with home refueling appliances, some units have experienced breakdowns. This drawback is likely to be corrected over time, as the technology is refined.

Local billing practices for natural gas use may also present a problem. Homes and businesses are generally equipped with a single meter that tabulates natural gas use. This use is generally billed at a residential or commercial customer rate. However, natural gas for transportation purposes is often charged at a lower "innovative market" rate, but may be subject to road taxes. Unless there is separate metering of the natural gas used by vehicle refueling appliances, proper billing may be impossible to achieve in some areas. Although the refueling appliance offers convenience, and separate measuring and billing is technically easy to accomplish, such metering could add another cost to an already expensive device.

Mobile Refueling

In addition to conventional refueling stations (see Profile 12) and refueling with a vehicle refueling appliance, there is a third type of natural gas refueling system. Several United States companies now offer mobile natural gas refueling services. Large compressed natural gas tanker trucks, periodically filled at a source of natural gas, travel among natural gas vehicle users for refueling. Each tanker truck carries approximately the natural gas equivalent of 1000 gallons of gasoline, enough to refuel about 70 natural gas vehicles. A tanker itself can be resupplied directly from natural gas distribution pipelines using on-board compressors. Tren-Fuels, Inc., an Austin, Texas-based subsidiary of Transco Energy Ventures Company, is one of the companies currently most active in the mobile refueling field. Although mobile refueling is not likely to be a long-term service as more permanent refueling options are developed, it now fills a need for commercial fleet operators who do not want to invest in their own refueling stations.

Profile 14 Removing Restrictions against Retail Natural Gas Sales: Federal and State Governments

One of the largest obstacles to the commercialization of natural gas vehicles and establishment of a refueling infrastructure to service them can be overcome at virtually no cost. Moreover, overcoming this obstacle does not require natural gas vehicle advocates to wait for improved technology or the economies of mass production. Indeed, overcoming this obstacle requires merely that legislators be convinced of the merits and advantages of natural gas vehicle use.

The obstacle in question consists of state and federal laws that prevent entities in the private sector from operating public natural gas refueling stations. These laws — so-called "sale-for-resale" prohibitions — were enacted decades ago, primarily to prevent price gouging by apartment building landlords. Before sale-for-resale prohibitions, landlords could pay for natural gas at rates set by state utility regulators and then resell the natural gas to individual apartment renters at exorbitant prices. The renters, lacking an alternative source of natural gas, were a captive market.

Sale-for-resale prohibitions work by declaring anyone wishing to resell natural gas to be a public utility. Thus, for purposes of these laws, a landlord or a natural gas vehicle refueling station operator is a public utility and must comply with detailed government regulations applying to public utilities, including extensive corporate reporting requirements, participation in lengthy natural gas rate approval and review procedures, and observance of rigid price controls.

Where sale-for-resale laws are in effect, they can thus stop the development of a natural gas refueling infrastructure dead in its tracks. This is true for both small and large entities that might wish to own and operate natural gas refueling stations. The cost of compliance with the regulatory requirements applying to public utilities exceeds the resources of small retail service station operators who might consider investing in natural gas refueling equipment. Even the management of a large company that might wish to invest in natural gas refueling — such as a large oil company — is likely to be deterred from doing so by the extensive financial public disclosure requirements of the utility regulatory process. Moreover, the requirement that public utilities sell natural gas at fixed, regulated prices would leave all companies incapable of reacting to price fluctuations in the markets for automotive fuels and at the mercy of price wars by their unregulated, gasoline-selling counterparts.

Sale-for-resale prohibitions have hampered natural gas vehicle demonstration programs, but have not entirely blocked them for two reasons. First, because their business already requires compliance with the sale-for-resale prohibitions, natural gas utilities can and do legally sell natural gas directly for vehicle use. Second, private companies that build natural gas refueling stations for use by their own vehicle fleets are similarly exempt. Because they are purchasing natural gas for their own use, no resale is involved and the prohibitions do not apply.

Removing Sales Restrictions

Widespread commercialization of natural gas vehicle technology, however, hinges on the establishment of an extensive natural gas refueling infrastructure. Since this is unlikely to happen without participation of private automotive fuel retailers, sale-for-resale prohibitions need to be expunged from state and federal laws.

Most regulation of natural gas sales to consumers is governed by state laws and, between 1989 and October 1992, eleven states removed sale-for-resale prohibitions: California, Colorado, Louisiana, Minnesota, New Mexico, New York, Oklahoma, Texas, Virginia, West Virginia, and Wisconsin. In most of these cases, removing the sale-for-resale prohibitions was relatively simple, requiring only the insertion of a clause into a statute's legal definition of the term "public utility" that provided an exemption for companies that buy natural gas for resale as an automotive fuel. In other cases, a ruling by the state's public utility commission has been sufficient to permit the resale of natural gas by private companies for transportation applications.

There are also federal laws pertaining to the sale to consumers of natural gas. The Federal Energy Regulatory Commission (FERC) regulates interstate natural gas sales; these sales typically involve natural gas pipeline transmissions. Under a narrow interpretation of the federal Natural Gas Act, which guides FERC's policies, the transmission of natural gas by interstate pipeline for use in natural gas vehicles could trigger a designation of the pipeline company as a "natural gas company" subject to full regulatory oversight by FERC.

FERC has interpreted existing law to permit the sale by pipeline companies of natural gas transported across state lines for use in natural gas vehicles as an unregulated business transaction subject to prevailing market conditions, not FERC control. In 1992, FERC initiated a rule-making procedure to formalize this position. Docket RM92-2-000 proposed to issue an exemption from FERC control, in the form of a "Generic Blank Certificate," for interstate commerce involving vehicular natural gas. In a July 16, 1992 decision, FERC officially exempted from its jurisdiction the sale of natural gas as a vehicular fuel.

Chapter 5 Government Incentives

The United States automotive and oil industries are both large and slow to change. Having spent the better part of a century perfecting gasoline-powered automobiles and the refueling infrastructure needed to keep them running, these industries have little incentive to radically alter the nature of their operations.

Automobiles are a central feature of life in the United States. More than 190 million vehicles are now on the road in the United States (just over one-third of the approximately 550 million vehicles in operation worldwide) — or an average of 1.8 automobiles per household. Average annual car and truck sales in the United States, including imports, top 15 million.

Collectively, United States residents spend about $370 billion per year to own and operate these cars. The annual cost of oil imports alone is $50 billion; another $83 billion is spent annually for imports of cars and parts. The average annual cost of transportation is $5220 per family, making it the second largest category of household expenses after the cost of shelter.

Moreover, the automotive and oil industries are economically important in terms of jobs. The United States produces more than 11 million cars and trucks each year, all powered by oil-derived fuels. Nearly 15 percent of total United States employment, or 12.3 million jobs, is related to vehicle production and use; for example, there are nearly 4000 motor vehicle and equipment manufacturing facilities in the United States, with a combined annual payroll of $35.6 billion. The Big Three auto makers together spend more than $200 million annually on advertising to promote this status quo.

The size and established infrastructure of the nation's automotive transportation system present formidable barriers to the introduction of technological innovations. Indeed, the automotive and oil industries have historically resisted some important technological changes. For example, the automobile industry spent millions of dollars to postpone introduction of the installation of seat belts. Another multimillion dollar effort now attempts to block congressional efforts to establish stricter automotive fuel efficiency standards. It is therefore not surprising that the changes required to effect a transition to natural gas vehicles, including major revamping of automotive design and the refueling infrastructure, are not being enthusiastically welcomed by the automotive and oil industries.

In the face of this resistance, government incentives for a transition to natural gas vehicle production and manufacture are needed. Several policy issues support such intervention. Chief among them are concerns about the adverse national security and environmental consequences of continued reliance on oil as the primary source of United States transportation fuel. These concerns increasingly point to the need for government to promote production, manufacture, and use of vehicles powered by alternative fuels.

Government intervention in the energy marketplace is nothing new; the oil industry, for one, grew as a result of government incentives. For instance, to encourage the exploration and development of oil wells, in 1918 the federal government offered favorable tax treatment to oil companies for investment losses due to intangible drilling costs and abandonment. This allowed the oil companies to recover 40 to 50 percent of their investment in dry holes (wells that do not produce oil) on their income taxes. In 1926, the government extended this preferential tax treatment by enacting the oil depletion allowance, which allows companies to deduct from their tax liability 27.5 percent of gross income earned from oil sales. By 1975, the oil depletion allowance, which has since been repealed, had cost the federal government more than $3 billion in tax revenue.

For the past decade, the ethanol industry has been a major beneficiary of generous government tax preferences. If ethanol is blended with gasoline to form a 10 percent ethanol fuel mixture (often called gasohol), the federal government allows an exemption from the highway tax of about $0.60 per gallon. With over 800 million gallons of ethanol being mixed with gasoline each year, this tax break represents foregone tax revenues of nearly $500 million annually.

The many advantages of natural gas as a transportation fuel argue in favor of creating comparable incentives for natural gas use or at least taking action to place natural gas on an equal footing with other transportation fuels. This chapter discusses five key types of incentives. The examples profiled in this chapter include government inducements favoring natural gas use, such as grants, loans, tax credits, tax exemptions, rebates, and favorable regulatory agency treatment, as well as government mandates to use natural gas, such as requirements that a certain number of vehicles employed for specific purposes be converted to natural gas use.

Government Incentives for Natural Gas Vehicles

What Needs To Be Done	Key Actors
15. Alternative fuel use mandates	Texas and other states
16. Grant programs	New Mexico and the Urban Consortium
17. Financing programs	Oklahoma and Utah
18. Preferential tax treatment	Oklahoma and California
19. Favorable regulatory climate for natural gas vehicle investments	California Public Utility Commission

Profile 15 Alternative Fuel Use Mandates: Texas and Other States

To create a large, steady market demand for alternative fuel vehicles, the United States Congress, 13 state legislatures (Arizona, California, Colorado, Florida, Iowa, Kansas, Louisiana, Missouri, New Mexico, New York, Oregon, Texas, and Washington), and the District of Columbia have enacted mandatory alternative transportation fuel use programs. The hope is that this demand will stimulate production of alternative fuel vehicles by the major automotive manufacturers and encourage the construction of appropriate refueling facilities. The government has mandated the use of alternative fuels in certain applications; however, all aspects of the production and sale of alternative fuel vehicles and refueling stations remain in the private sector.

Each of these programs mandating alternative fuel use targets specific vehicle fleets for conversion and establishes a schedule to be followed in implementing the fuel switch. Although these programs are fuel-neutral (that is, the mandates can be fulfilled through the use of any of a number of alternative transportation fuels, including methanol, ethanol, and propane), natural gas is the front-runner in most cases. The federal alternative fuel mandate was discussed in Chapter 1 and Profile 10. State programs are the focus of this profile.

The Texas Program

In June 1989, Texas became the first state to enact a comprehensive law mandating the use of alternative fuels. The federal government and several other states have followed suit, but the Texas program remains the most ambitious mandatory alternative fuel use program in the country.

Two complementary statutes provide the backbone for the Texas program. The law emanating from Senate Bill (SB) 740 requires alternative fuel use in vehicles owned by state government agencies, public school districts, and metropolitan and regional transit authorities. The statute created by SB 769 applies to local government and privately owned fleet vehicles operated in the 21 Texas counties – containing about 70 percent of the state's population – in which air quality violates federal public health standards ("non-attainment" counties). Both statutes define alternative transportation fuel to include compressed natural gas or any other fuel that produces comparable or lower air pollutant emissions. To date, the Texas Air Control Board has also certified propane, methanol, and electric vehicles as other qualifying alternative fuel vehicles.

The mandate of SB 740 affects all state government fleets containing more than 15 vehicles, school bus fleets larger than 50 buses, and all metropolitan and city transit departments, regardless of the size of their fleets. Since September 1, 1991, fleets covered by the bill have been required by SB 740 to purchase or lease vehicles capable of running solely on alternative transportation fuels. These vehicles may be factory-built to burn only alternative fuels, or they may be conventional vehicles converted to "bi-fuel" use; that is, able to run on either gasoline or alternative fuels.

Fleets covered by the bill are affected by the mandate of SB 740 even if they have no plan to purchase or lease new vehicles; that is, they are required to convert existing vehicles to alternative fuel use. By September 1, 1994, SB 740 requires that at least 30 percent of the fleet vehicles covered by the bill be capable of burning alternative fuels, and by September 1, 1996, 50 percent of public school buses, municipal buses, and state agency vehicles must be capable of burning alternative fuels. By September 1, 1998, at least 90 percent of all fleets covered by the bill could be required to convert to alternative fuel use, if an interim review by the Texas Air Control Board confirms that the program is effective in reducing air pollution.

Most of the provisions of the second bill, SB 769, will go into effect only if SB 740 does not reduce pollution sufficiently. SB 769 requires the Texas Air Control Board to study the impact of the alternative fuel mandate program established in SB 740 and, in 1996, to publish a report of its findings. If the report concludes that SB 740 has been effective in reducing pollution, but that its requirements alone are insufficient to ensure that air quality in the 21 non-attainment Texas counties meets federal public health standards, then the additional alternative fuel use mandates of SB 769 come into play.

If the supplemental mandates of SB 769 are triggered, they could potentially be extended to nearly all vehicles owned by local government and private sector fleets. SB 769 would require that at least 30 percent of all local government fleets containing more than 15 vehicles and private sector fleets of more than 25 vehicles in the 21 non-attainment counties be capable of burning an alternative fuel by September 1, 1998. This percentage rises to 50 percent by September 1, 2000, and to 90 percent by September 1, 2002.

The Texas program has had a dramatic impact on the conversion of vehicles to natural gas use. Scores of fleet conversions are currently underway in Texas, and dozens of natural gas refueling stations are under construction. The Texas Railroad Commission has established a certification program under which equipment marketed by natural gas conversion companies is licensed for sale in the state, and a toll-free phone line operated by the Texas General Land Office (1-800-6-FUEL-99) provides callers with answers to questions about the alternative fuel program. In April 1992, the third annual Texas Alternative Fuel Conference drew 1600 participants, making it the largest event ever held to promote alternative fuel technology. (See Profile 24 on other alternative fuels conferences.)

Other State Programs

In addition to the Texas and federal programs, the 13 states listed at the beginning of this profile have established alternative transportation fuel programs mandating alternative fuel use in designated fleets, some of which are summarized here.

- A 1989 California law requires that 25 percent of all newly acquired state government vehicles be capable of burning alternative fuels, subject to the availability of alternative fuel vehicles. In addition, a 1991 California law requires that all vehicles-for-hire use alternative fuels if they operate in non-attainment zones.

- A 1990 Colorado law requires that 10 percent of all new vehicles purchased or leased by the state government during the 1991-1992 fiscal year operate on alternative fuels. This percentage rises an additional 10 percent each year until the 1994-1995 fiscal year; thereafter, 40 percent of all new state government vehicle purchases or leases must continue to be of vehicles that operate on alternative fuels.
- Beginning in 1998, a 1990 District of Columbia law prohibits the operation of commercial vehicles not powered by alternative fuels during daylight hours from May 1 through September 15. The District's law also requires operators of fleets containing 10 or more vehicles to convert 5 percent of their fleets to operate on alternative fuels by 1993. This percentage increases an additional 5 percent each year through the year 2000.
- A 1990 Louisiana law requires alternative fuel use in 30 percent of new state government vehicles by September 1, 1994. The percentage could increase to 80 percent in 1998, pending an analysis of the program's effectiveness by the state's Department of Environmental Quality.
- In 1991, Washington legislators passed a law requiring 30 percent of new state agency vehicles to use alternative fuels after July 1, 1992. This percentage increases by 5 percent each year.
- In March 1992, the New Mexico legislature enacted the Alternative Fuel Conversion Act which requires that, starting in mid-1993, 30 percent of new state vehicle purchases be capable of burning alternative transportation fuels. The percentage rises to 60 percent in mid-1994 and to 100 percent in mid-1995.

Profile 16 Grant Programs: New Mexico and the Urban Consortium

Alternative fuel use mandates (discussed in Profile 15) are undoubtedly the biggest "sticks" that a government can wield to promote a switch from reliance on oil-derived fuels for transportation. By contrast, direct monetary awards represent large "carrots" available to a government wishing to promote this switch. In fact, since 1990, state governments have greatly expanded grant programs aimed at encouraging the use of alternative transportation fuels: New Mexico was one of the first to do so, and its program provides a sizable amount of money on both a gross and per capita basis. In addition, local governments have initiated similar programs; five of these programs are coordinated by the Urban Consortium, an arm of the National League of Cities funded by the US Department of Energy.

The New Mexico Program

In 1989, New Mexico became one of the first states to offer significant grants for natural gas vehicle projects when the state's Energy, Minerals and Natural Resources Department (EMNRD) initiated a $2.1 million transportation grant program. As of October 1992, the New Mexico EMNRD had awarded approximately $1.8 million to fund 176 natural gas vehicle conversions. Other government and private sector project sponsors had contributed an additional $1.5 million, for a total commitment of over $3.0 million. The largest of these six grants, for $500,000, funded the purchase and operation by the City of Roswell of two natural gas buses produced by Transportation Manufacturing Corporation (TMC). Located in Roswell, TMC is the nation's largest producer of full-size urban buses. Unlike its major competitors, Flxible Corporation and Bus Industries of America, TMC focused its attention in the late 1980s on production of methanol-powered buses for sale in California. The EMNRD grant could help TMC enter the natural gas bus market while achieving its main purpose of helping Roswell establish its first mass transportation system. An additional grant will permit Roswell to supplement its two natural gas buses with five natural gas-powered minivans and three conventional diesel-powered buses.

A second EMNRD effort funds a pilot natural gas school bus project. Initiated in mid-1991, the project is being conducted jointly with New Mexico's Department of Education. Its objective is to convert approximately 100 school buses to natural gas use. Grant awards of about $5000 per bus are being offered to finance the bus conversions. The first such award, for conversion of 30 buses, was issued to the Los Lunas School District. The school buses were converted in 1992, and New Mexico Governor Bruce King cut the ribbon at the official opening of the refueling station for these buses on August 12, 1992. In September 1992, projects in two other cities, Belen and Las Vegas, were approved. The state's largest natural gas utility, Gas Company of New Mexico, is contributing the refueling stations for most of these projects.

In 1991, Santa Fe Community College was the recipient of a $150,000 EMNRD grant to fund the development of college-level curricula to be used for instructing auto mechanics about all types of natural gas vehicles. The grant will also cover the cost of converting a 30-passenger natural gas-powered bus for use in the campus shuttle service. In the fall of 1992, the course resulting from this grant was taught for the first time.

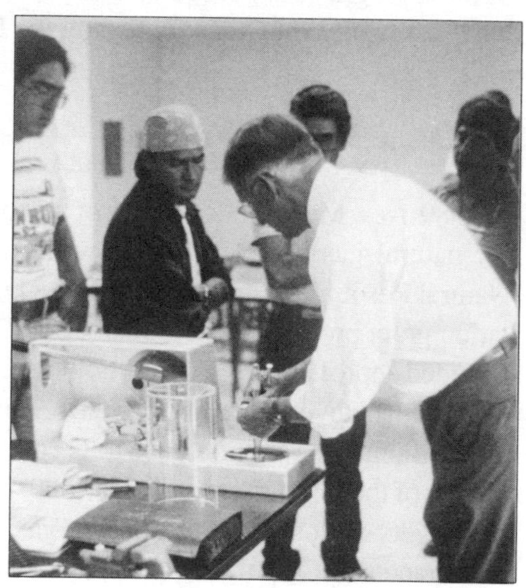

Students learn the mechanics of natural gas vehicles at Sante Fe Community College. Photo: Sante Fe Community College

The other three EMNRD projects involve individual

grants to municipalities for natural gas vehicle conversions. The grants include 30 animal control vans — the local "dog-catcher" — owned by the City of Albuquerque, 30 pickup trucks owned by the City of Roswell, and four vans operated by the Chaves County Retired Senior Volunteer Program. In addition, EMNRD converted five of its own vehicles to natural gas use in the spring of 1992 and has installed two slow-fill refueling pumps at its headquarters.

In 1992, the EMNRD and the Gas Company of New Mexico received a President's Environmental Challenge Award for their effort to promote natural gas vehicle technology.

Other State Programs

The New Mexico alternative fuels program was among the earliest such programs concentrating on natural gas vehicle development. On a per capita basis, it remains one of the largest. However, many other states have followed New Mexico's example, including the two cited below.

The most ambitious program started in the 1990s is probably New York's Alternative Fuels for Vehicles Demonstration Program. New York's program is an outgrowth of nearly a decade of research funded by the New York State Energy Research and Development Authority (NYSERDA). During that period, NYSERDA funded some natural gas vehicle projects, but most of its efforts involved methanol vehicle demonstrations and use of biomass fuels such as wood from the abundant forests in the state.

In August 1990, New York Governor Mario Cuomo announced that NYSERDA would coordinate a new multi-agency task force called the Alternative Fuels Coordinating Council. The council, staffed and administered by NYSERDA, has developed the 15-year Alternative Fuels for Vehicles Demonstration Program. More than 250 alternative fuel vehicles will eventually be included in this program. Funding for the first five years, through 1994, is set at $38 million, making this one of the largest state government-directed undertakings in the field of alternative transportation fuels. Half of the $38 million will come from the state treasury, with the Federal Transit Administration and the Department of Energy contributing much of the remainder; some funding will also come from the private sector. Natural gas vehicles will constitute the largest category of alternative fuel vehicles under this program.

Pennsylvania's Alternative Transportation Fuel Program, established in 1988 by the State Energy Office, is also especially noteworthy because of its size and its emphasis on natural gas use. In 1989, the Energy Office initiated a $2 million funding program as part of a larger state program called "Oil Import Limits in Pennsylvania"; more than 30 alternative transportation fuel projects were funded through mid-1992. Efforts promoting natural gas vehicle technology are at the heart of this undertaking, accounting for about 40 percent of the program's proposed expenditures. Natural gas projects funded by the program through mid-1992 have included the conversion of 125 vehicles by the State Department of Transportation and conversion of about 200 other fleet vehicles and buses, mainly in the urban centers of Philadelphia and Pittsburgh.

Urban Consortium Municipal Initiatives

Independent of state government-sponsored activities, municipalities are also getting involved in alternative transportation fuel projects. For instance, five municipal projects coordinated by the Urban Consortium Energy Task Force together constitute a major effort to promote natural gas vehicle use in United States cities. Under the general oversight of the National League of Cities, the Urban Consortium Energy Task Force includes representatives from over 40 of the largest United States cities and most densely populated counties. The Task Force, established in the early 1970s, funds and coordinates municipal projects to apply new technology and advanced management systems within city governments. Money for the projects conducted by member cities of the Urban Consortium is provided by the US Department of Energy.

In the late 1980s, the Energy Task Force established an "Alternative Vehicle Fuels and Technologies Unit." In 1992, the Urban Consortium funded five projects involving natural gas vehicles as part of the work of the Energy Task Force:

- A study of the barriers to implementing large-scale fleet natural gas vehicle utilization in non-attainment areas, being conducted in Houston, Texas.
- Development of a natural gas refueling infrastructure in Pittsburgh, Pennsylvania.
- A demonstration project involving natural gas, hythane, electric, and hybrid vehicle technology in Denver, Colorado (see Profile 9). The project also includes development of an alternative fuel evaluation system that can assist cities in the design and implementation of alternative transportation fuel programs.
- Development of a multi-fuel service station in Albuquerque, New Mexico.
- Preparation of an analysis of a compliance strategy under which the non-emergency vehicle fleet in San Diego, California can meet the requirements of state and federal air pollution regulations.

Profile 17 Financing Programs: Oklahoma and Utah

Financing for natural gas conversion projects is another "carrot" available to government—an incentive that can help reduce the barriers to natural gas use posed by the capital costs of switching to an alternative transportation fuel. Low-interest or no-interest loans to pay the costs of converting a vehicle to burn natural gas can significantly lessen this economic burden. By increasing the number of vehicle owners who can afford to convert to natural gas use, government financing programs can, if properly structured, encourage development of a stronger market for natural gas vehicle technology. As of October 1992, 13 states had financing incentive programs such as those described in this and the following profile: Arizona, California, Colorado, Connecticut, Louisiana, New Mexico, Oklahoma, Oregon, Pennsylvania, Texas, Utah, Washington, and Wisconsin.

Although the average life cycle of a heavily driven natural gas vehicle makes it cheaper to own and operate than a gasoline-powered vehicle, the capital cost of conversion is at present a major economic hindrance to the commercialization of natural gas vehicle technology. Conversion kits suitable for most gasoline-powered automobiles cost between $2500 and $3000, including installation. Medium-duty engine conversions cost between $5000 and $10,000, while the cost of converting a large urban transit bus can exceed $30,000.

Assuming that a converted natural gas vehicle is driven at least 20,000 miles per year, the capital cost of conversion is usually repaid in less than five years because of the fuel price advantage natural gas enjoys by comparison to gasoline (see Chapter 1). Moreover, the second generation of natural gas vehicle conversions entails lower capital expenditures because many of the parts installed in a natural gas conversion, including the storage cylinders that account for about one-third of the total conversion cost, can be reinstalled on other vehicles after the first vehicle converted is scrapped.

Mass production of dedicated natural gas vehicles will eventually further improve the comparative economic advantages of natural gas vehicles. A 1990 report by the United States Alternative Fuels Council, for instance, is one of several studies predicting that, assuming current technology, an assembly line-produced natural gas vehicle could, by the mid-1990s, cost only about $800 more than its gasoline-powered counterpart, or less than a third of the current conversion cost. Thus, once assembly line natural gas vehicle production is firmly in place, the low price of natural gas will make natural gas vehicles much more cost-effective to own and operate than gasoline-powered vehicles.

State Financing Programs

Oklahoma and Utah were the first two states to have financing programs in place for alternative fuel vehicles and, as of October 1992, these were the only programs with money actually available for loans. The first and to date the larger of these is the Oklahoma Alternative Fuels Conversion Fund, established in 1990. Created by the Oklahoma Alternative Fuels Conversion Act (HB-2169), which was signed into law by Governor Bellmon on May 31, 1990, the revolving loan fund was initially capitalized at $1 million, but has proved so popular that an additional $500,000 was added to the fund in 1991. In addition, the program's reach was expanded in 1991 to include loans for financing the construction of alternative fuels refueling stations, as well as for the vehicle conversions covered in the original effort.

Oklahoma's program offers seven-year, no-interest loans of not more than $3500 per vehicle or $100,000 per refueling station to state, municipal, and county governments, and to school districts. Loans are approved by the Oklahoma Corporation Commission. The loans for vehicle conversions are repaid from fuel savings derived from alternative fuel use. It is the responsibility of the Oklahoma Tax Commission to calculate the fuel savings achieved by each loan recipient and to make periodic readjustments as the price of fuels fluctuates.

Although loans can be made for conversions to use of a variety of alternative fuels, applicants must be able to demonstrate the ability to repay the loans from fuel savings. Thus, the Oklahoma program is largely restricted to natural gas conversions, since natural gas is the only alternative fuel that is consistently cheaper than gasoline. Under the program, loan repayments would be suspended if the price of alternative fuels ever equaled or exceeded the price of gasoline. The government, therefore, assumes the risk that the price of alternative fuels (mainly natural gas) will remain lower than gasoline prices, making the loans risk-free to vehicle owners with regard to fuel prices.

As of October 1992, $1.2 million had been loaned to convert 152 vehicles and finance four refueling stations. The governor's limousine is one highly visible natural gas conversion funded by the Oklahoma program. More than $125,000 has already been saved through fuel savings from all natural gas conversions. This money is once again available for new loans.

School bus projects dominate the conversions financed by the Oklahoma loan program. The two largest school bus conversion projects include 40 buses owned by the Tulsa School District and 24 buses owned by the Enid School District; these conversions equipped the buses to use natural gas. The fuel savings from the Tulsa and Enid conversions were approximately $1000 per bus in the first year alone. The Tulsa School District plans to operate a fleet of 118 natural gas buses by the 1993 school year, including 24 buses that were converted to natural gas use as the result of a 1988 grant from the Oklahoma Department of Commerce.

Oklahoma's revolving loan fund is but one of several programs in the state designed to promote natural gas vehicle technology. Another Oklahoma law establishes tax credits for conversions to alternative fuel use (see Profile 18), and yet a third Oklahoma statute sets certification standards for the vehicle conversion industry.

Together, these laws have spurred an explosive growth in and attracted business to Oklahoma for natural gas vehicle conversions and construction of refueling stations. For example, as of October 1992, there were 22 refueling stations in Oklahoma and more were under construction. Tulsa-based Crane Carrier Corporation began selling natural gas truck chassis in 1991; it also markets natural gas-powered fork lift trucks, and a subsidiary has opened the Cleanaire Conversion Center, where natural gas conversion equipment is installed and tested. There are five other conversion centers in the state, all opened since 1989. Tri-Fuels Inc., of Baton Rouge, Louisiana, has relocated its natural gas compressor packaging operation to Oklahoma. In February 1992, United Parcel Service (UPS) announced a plan to convert 140 trucks it operates in Oklahoma to natural gas. This is the fifth state in which UPS has undertaken a natural gas conversion effort, and it is the company's largest program by far (see Profile 4). Oklahoma Natural Gas Company operates a fleet of 200 natural gas vehicles, and Tinker Air Force Base plans to convert 300 military vehicles to natural gas.

In May 1991, Utah followed Oklahoma's example and enacted two fuel incentive laws creating two revolving loan funds, one for public and one for private sector fleet vehicle conversions. Utah's Division of Energy administers the funds which provide loans of up to $3000 per vehicle for the conversion to alternative fuel use of state and local government fleet vehicles, as well as privately-owned and operated fleet vehicles. Repayment is required within seven years; loans accrue interest at the annual compounded rate of seven percent.

Utah's public and private revolving loan funds were activated on July 1, 1991. During the first year, each fund was capitalized at only $10,000. Thus, the funds did not provide a significant boost to vehicle conversions. In 1992, however, the Utah legislature increased each of the loan funds by $160,000.

In March 1992, the New Mexico legislature established a loan program, as part of the Alternative Fuel Conversion Act, to finance conversions to alternative transportation fuels. Although the law established a cap of $5 million for the fund, no money was appropriated to capitalize it during its first year. However, as of October 1992, the administering agency, the General Services Department, was proceeding to establish regulations to implement the financing program if and when money is appropriated for this purpose.

Profile 18 Preferential Tax Treatment: Oklahoma and California

Preferential tax treatment, including tax credits, tax exemptions, rebates, and lower tax rates, is another tool available to a government wishing to provide economic assistance to purchasers and operators of natural gas vehicles. These measures, which work by reducing an individual's or an entity's tax obligation, can be implemented without significant financial or administrative commitments and require comparatively little government oversight, compared, for example, to loan programs. Tax credits or deductions allow designated entities to reduce their annual income tax payments, while tax exemptions eliminate the requirement that certain taxes be paid at the time designated purchases are made. With rebates, a purchaser recovers from tax revenues or in a cash refund a portion of the purchase price of designated items. Finally, selectively low tax rates are another way to encourage certain economic activities.

Worldwide, each of the tax incentives mentioned above has been used to spur investment in natural gas vehicle technology. In New Zealand, for example, in the early 1980s, the government offered rebates of NZ$200 to the purchasers of natural gas conversion kits as part of its effort to encourage natural gas vehicle use. By 1986, this and other incentives had helped put more than 100,000 natural gas vehicles on New Zealand roads. As a result, when the incentive program peaked in 1988, natural gas vehicles constituted 11 percent of the automobile population in New Zealand — the highest rate of natural gas vehicle use in the world.

Since the early 1980s, a keystone of Canada's natural gas vehicle program has been a federal government rebate of C$500 for each vehicle converted to natural gas vehicle use. Furthermore, in the province of Ontario, sales of natural gas vehicle conversion kits are exempt from the 8 percent provincial sales tax, and a sales tax exemption of up to C$1000 is offered purchasers of new automobiles converted to burn natural gas immediately after purchase.

In the United States, Oklahoma and California have the largest tax credit initiatives in place for automobile natural gas use. Colorado, Connecticut, Utah, and New York have initiated similar systems, and the new National Energy Policy legislation offers tax benefits on the federal level. In addition, road taxes provide another option for tax preferences for natural gas vehicle use.

Oklahoma and California

On May 31, 1990, the first tax credit initiative for automobile natural gas use was established in Oklahoma as a result of a law, HB-2169. Beginning on January 1, 1991, this law offered tax credits for private sector natural gas vehicle conversions. The credit is for 20 percent of the actual cost of the conversion equipment or the refueling station equipment (such as a compressor). Furthermore, HB-2169 established a 10 percent tax credit, up to a maximum of $1500 per vehicle, for the total purchase price of a vehicle initially built to burn only natural gas.

Oklahoma's tax credit program had barely begun when, on May 24, 1991, as an added incentive to perform natural gas vehicle conversions, the state legislature raised the amount of the tax credit. The amended law raised the tax credit from 20 to 50 percent for the two-year period ending January 1, 1993, at which time the amount of the credit would revert to 20 percent. The amended law (contained in HB-1193) also expands the fuel conversions that qualify for the tax credit to include conversions to liquefied petroleum gas. Conversions to other alternative transportation fuels, however, such as ethanol, methanol, and electricity, remain ineligible for the tax credit.

California's preferential tax treatment provisions are contained in two complementary laws. The first piece of legislation (SB 1006) offers a sales tax exemption. Enacted in 1989, it makes the California Air Resources Board responsible for designating vehicles that qualify as low-emission vehicles (meaning vehicles that meet pollution standards, set by the Board, that are more stringent than federal requirements), including those running on alternative fuels. (California's four levels of pollution standards are described in detail in Profile 20.) The law then directs the California Energy

Commission to determine the incremental cost of purchasing low-emission vehicles by comparison to the cost of purchasing conventional vehicles that meet current federal and state air pollution standards. Purchasers of designated low-emission vehicles or alternative fuel conversion equipment are exempt from paying sales tax on the incremental cost. This tax exemption is effective through December 31, 1994.

The second preferential California tax statute (SB 2600), enacted in September 1990, offers a tax credit, rather than a sales tax exemption, to purchasers of low-emission vehicles or equipment needed to convert conventional vehicles into low-emission vehicles. This law allows qualified purchasers to deduct from their income taxes an amount equal to 55 percent of the incremental cost of the low-emission vehicle. These credits, however, are limited to a total of $1000 per passenger automobile and $3500 for every other kind of vehicle. These credits can also be claimed through December 31, 1994.

Other States and the Federal Government

Four other states currently have similar tax credits. Connecticut Act 91-179 established a 10 percent investment tax credit for expenditures related to alternative transportation fuels. In Colorado, the state government offers a $200 cash rebate to purchasers of new alternative fuel vehicles or those who convert an existing vehicle to run on alternative fuels. There is a limit of five rebates per applicant. A law enacted in New York in July 1992 provides tax breaks for alternative fuel vehicles, exempting conversion equipment from state taxes for five years. The tax relief is expected to range from $150 to $300 per vehicle converted to natural gas. In mid-1992, Utah enacted a $400 tax credit for the purchase of new or converted alternative fuel vehicles.

Finally, the October 1992 National Energy Policy Act establishes the first federal tax incentive for alternative fuel vehicles. It permits builders of refueling stations for alternative transportation fuels to deduct up to $100,000 per year from their federal tax calculations for investments in each refueling station. Similarly, investments in the incremental cost of vehicles running on alternative transportation fuels can be deducted: up to $2000 for light-duty cars and trucks and $10,000 for heavy-duty trucks and buses.

Road Taxes

Another tax preference favoring the use of natural gas as a transportation fuel exists not so much by design as by oversight: natural gas is exempt from federal road taxes and from most state road taxes. For decades, road taxes have been collected and used to fund road construction and maintenance. The current federal road tax, which applies to petroleum-derived fuels, is $0.14 per gallon and the average state road tax is $0.18 per gallon. The highest state road tax is $0.26 per gallon charged by Rhode Island; the lowest is $0.07 per gallon charged by Georgia. With over 100 billion gallons of petroleum-derived automotive fuel sold each year, road taxes annually raise more than $30 billion in government revenues.

In most states, the favorable road tax rates for natural gas use exist partly as an incentive to encourage use of alternative fuels. This incentive also exists because of the difficulty and high cost of precisely measuring natural gas dispensed at refueling stations; there is no uniformly accepted conversion ratio to translate cubic feet of natural gas to an equivalent volume of liquid gasoline for taxation purposes.

Nine states (California, Colorado, Indiana, Minnesota, New Mexico, Oklahoma, Tennessee, Utah, and Washington) have, however, replaced the standard road tax rate, based on cents per gallon, with a permit system applicable to natural gas vehicles. For example, in New Mexico, a natural gas vehicle owner is required to pay, in lieu of road taxes, $75 annually for a permit to operate the vehicle. In California, Colorado, and Oklahoma, the annual natural gas vehicle permit fees for passenger automobiles are $36, $70, and $100, respectively.

A few states have also adopted conversion ratios that translate natural gas usage to an equivalent amount of gasoline usage and then set road tax rates accordingly. In New York, for example, a road tax of $0.08 is collected for each equivalent gallon of natural gas used as a transportation fuel — the same rate that is collected for gasoline use. In Iowa, natural gas is taxed at $0.16 per equivalent gallon, 20 percent below the $0.20 per gallon rate for gasoline.

The question of whether road taxes should be applied to natural gas use is likely to garner increasing attention in future years. In the meantime, at least, natural gas vehicle users generally benefit from no road taxes or lower road taxes than those paid by users of gasoline-powered vehicles.

Profile 19 Favorable Regulatory Climate for Natural Gas Vehicle Investments: The California Public Utility Commission

Natural gas is among the most heavily regulated of all commodities sold in the United States. The regulatory structure for natural gas was set in place by enactment of the 1935 Public Utilities Holding Company Act and by subsequent legislation. Since then, the activities of producers, distributors, and sellers of natural gas have been closely monitored by federal and state utility regulators. Although the federal government regulates interstate sales of natural gas, mainly through the Federal Energy Regulatory Commission (FERC), the activities of most natural gas companies fall under the jurisdiction of state agencies, usually called public utility or public service commissions (PUCs or PSCs). The way in which the regulatory posture of these agencies evolves with respect to using natural gas as a transportation fuel has tremendous bearing on the future of natural gas vehicle technology.

The "gospel" of United States utility regulation begins with the assumption that natural gas companies must make capital investments to deliver natural gas to the point of sale. The sum of reasonable capital investments makes up what is called the rate base. Regulators allow utilities to recover rate base investments, plus the cost of the natural gas itself, operating expenses, and a fair profit, from individual customers, or ratepayers. Public utility commissions determine what constitutes "reasonable" capital and operating costs and "fair profits" on a year-to-year basis.

Each year, the size of the rate base changes to reflect additional investments in customer natural gas "hook-ups," other capital expenses, and equipment depreciation. New customers are not charged the full cost of their individual hook-ups (the links that deliver gas from the pipeline to a home or business); instead, hook-ups and related expenses are added to the rate base and averaged into the bills of all customers. Expenses included in the rate base generally include all equipment up to and including an individual customer's natural gas meter. After the meter, however, equipment such as piping linking the meter to various appliances is a private expense, the sole financial responsibility of the customer.

Deciding when investment in natural gas vehicle technology becomes a private expense, rather than an expense borne by the utility, is a key issue in the context of future regulation of natural gas. For instance, if regulators conclude that the point at which a natural gas distribution pipeline enters a refueling station is the point at which costs become private rather than utility expenses, neither the costs of refueling stations nor the costs of equipment required to convert a vehicle to natural gas use can be shared by all natural gas users after being figured into the rate base. By contrast, if regulators conclude that the individual customer's financial responsibility begins when natural gas enters a natural gas vehicle engine, then all costs of refueling station construction and vehicle conversion can potentially be included as utility expenses in the rate base, thus minimizing the cost of individual refueling stations or natural gas vehicles. These decisions do not have to be at either extreme: for instance, a middle ground would be a decision to include in the rate base only the cost of natural gas compressors and fuel dispensers at refueling stations.

In addition to decisions regarding inclusion of capital investments in the rate base, utility commissions have other regulatory tools with which they can either promote or hinder the development of natural gas vehicle technology. They can, for example, authorize "incentive rates" – rates for the sale of natural gas as a transportation fuel that are lower than the actual cost of producing and delivering the fuel to the end user. In this way, part of the cost of natural gas as a transportation fuel is borne by all the natural gas ratepayers in the area, whether they use natural gas vehicles or not. The argument for such a rate structure is that everybody benefits from increased natural gas vehicle use (through lesser air pollution, for example).

Regulators can also tie natural gas rates to the prevailing price of gasoline in a way that maximizes (or minimizes) the fuel price advantage for natural gas. They can authorize utilities to engage in natural gas vehicle conversion equipment sales or to make natural gas vehicle equipment leasing arrangements with refueling station operators. They can limit or encourage advertising campaigns endorsing natural gas vehicle use, as well as the establishment of unregulated subsidiaries or affiliates organized to participate in natural gas vehicle development activities. In short, the power of utility regulators to shape the future of natural gas vehicle use is considerable.

California's Regulatory Actions

Perhaps more than any other state, California has grappled with the issues concerning utility commission regulation of natural gas vehicles, especially the issue of what capital investments to include in the rate base. California has created a regulatory structure that effectively encourages the use of natural gas vehicles. Efforts by the California Public Utility Commission (CPUC) were in large part propelled by a 1990 law (SB 2103) directing the CPUC to "evaluate and implement policies to promote the development of equipment and infrastructure needed to facilitate the use of... low-emission vehicles," specifically natural gas and electric vehicles. The CPUC is required to coordinate its activities with other state agencies and the automotive industry, and must submit progress reports to the legislature every two years beginning January 30, 1993.

The CPUC evaluation process was almost immediately put to a test as the result of applications from two of the state's largest utility companies for rate-setting decisions regarding proposed major natural gas vehicle programs. San Diego Gas and Electric Company applied on June 11, 1990, to expand its natural gas vehicle program and to include a major portion of its program expenses in the rate base. A similar application was filed by Pacific Gas and Electric Company on July 26, 1990.

The CPUC ruled on both applications on July 2, 1991, and authorized program expenses for at least a two-year period. In the case of San Diego Gas and Electric Company, the CPUC approved expenditures of $6.8 million over a two-year period. The program includes construction of 12 new natural gas refueling stations and a $500,000 conversion rebate fund for individual vehicle conversions. Each rebate may not exceed 50 percent of the total vehicle conversion cost, however.

The ruling on the application from Pacific Gas and Electric Company similarly supported its natural gas vehicle programs. Pacific Gas and Electric's total program will cost $18 million. With that money, the utility will build six refueling stations for its own use, six public refueling stations built jointly with oil companies, and 13 privately owned and operated stations. The CPUC order also authorizes the utility to offer rebates of up to $1250 to private fleet operators for each vehicle they convert to natural gas, up to a maximum of 50 percent of the total fleet conversion cost. Other approved activities include a natural gas vehicle marketing program, hiring and training of technical support personnel, a conversion program for utility-owned vehicles, and the expansion of data collection and analysis efforts concerning natural gas vehicles.

The CPUC rulings also established incentive rates for natural gas sales by both utilities. In the San Diego Gas and Electric Company service area, the price of compressed natural gas as a transportation fuel was set at about $0.50 per equivalent gallon of gasoline for buses and $0.70 per equivalent gallon for non-bus uses. Compressed natural gas may be sold as a transportation fuel by Pacific Gas and Electric Company for about $0.62 per equivalent gallon in addition to a $11.99 monthly customer service charge. In both cases, lower rates were set for the sale of uncompressed natural gas to refueling stations equipped with their own compression equipment. While these prices would certainly be higher without these incentive rates, natural gas rate structures are so complex, with different rates for many different customer groups, that there is no single rule of thumb for determining precisely what "non-incentive" rates would be.

A third application was filed before the CPUC on July 1, 1991, by Southern California Gas Company. This proposal was more extensive than the previous two proposals combined. Southern California Gas, which services the Los Angeles area, planned to spend $43 million over a two-and-a-half-year period to build 51 natural gas refueling stations; it also hoped to offer rebates for natural gas vehicle conversion and for the purchase of natural gas vehicles produced directly by automotive manufacturers. On January 10, 1992, the CPUC approved a portion of the Southern California Gas Company program, consisting of expenditures totaling $10.8 million, to develop and market natural gas refueling stations and vehicles over the next two years.

Chapter 6 Removal of Institutional and Regulatory Barriers

Just a few years ago, serious promotion of natural gas vehicle technology in the United States was virtually nonexistent. Today, however, a technological revolution in automotive transportation is taking place at an astonishing speed. For example, since 1990 nearly all major United States manufacturers of automobiles, trucks, and buses have developed at least prototypes of natural gas vehicles, and several are now offering fully warranted products for commercial sale. Nonetheless, the automotive status quo remains structured to support the use and development of gasoline-powered vehicles. Significant institutional barriers exist to the advancement of alternative fuel technologies in general and to natural gas vehicle research, development, and use in particular.

The structural basis of the automotive status quo is legal and regulatory: a complex set of laws, regulations, standards, and ordinances have been passed to address issues raised by the use of gasoline-powered vehicles, sometimes inadvertently to the exclusion of natural gas. However, the automotive status quo is supported by more than just laws and regulations. Public perceptions play a major role, too. The public has intimate familiarity with conventional automotive technology, acquired through a lifetime of daily use. Also, a vast body of literature about conventional vehicles — professional journals, popular magazines, brochures — is produced to satisfy any lingering curiosities of readers ranging from members of the Society of Automotive Engineers to home mechanics and consumers who visit automotive showrooms. Public experience with and knowledge of natural gas vehicle technology, on the other hand, is very sparse. As a result, commonly held public opinions about natural gas vehicles, where they exist, are as frequently based on misconceptions as on accurate information.

This chapter discusses six actions that aim to remove or chip away at the major institutional barriers to natural gas vehicle use. In order for natural gas vehicle technology to become rooted in United States culture as an accepted heir to gasoline-powered vehicles, public awareness and understanding must be greatly expanded. In addition, a comprehensive regulatory framework must be established that is specifically tailored for natural gas vehicles but that complements current laws and regulations affecting conventional vehicle use. In order to be durable and stable, the rules concerning natural gas vehicle production, manufacture, and use must be supported by accurate technical information and reflect sound decision-making. Answers to questions that are still perplexing today must be sought, found, and incorporated into policies and regulations. Finally, international cooperation is vital to efficient development of natural gas vehicle technology, and all aspects of this new industry need to be monitored over time and reported to audiences worldwide on an ongoing basis.

Removing Institutional Barriers to Natural Gas Vehicles

What Needs To Be Done	Key Actors
20. Setting emissions and safety standards	US Environmental Protection Agency and California Air Resources Board
21. Removing local natural gas vehicle use restrictions	New York City
22. Investigating unanswered environmental issues	International Energy Agency
23. Providing a knowledgeable voice for natural gas vehicle interests	Natural Gas Vehicle Coalition
24. Promoting public education about natural gas vehicles	Specialized conferences and publications
25. Supporting international information exchange and cooperation	International Association for Natural Gas Vehicles

Profile 20 Setting Emissions and Safety Standards: United States Environmental Protection Agency and the California Air Resources Board

Dozens of federal and state laws and thousands of pages of regulations affect automotive production and use in the United States. To a great extent, these laws and regulations can be applied to natural gas vehicles as easily as to gasoline-powered vehicles. For example, existing seat belt requirements can be applied equally well to both types of vehicles. In two crucial areas, however, natural gas vehicle technology is different enough so as to warrant special regulatory treatment: environmental and safety controls.

Two examples of areas in which regulations will be needed to respond to the special environmental and safety concerns associated with natural gas vehicles are emissions standards and cylinder construction standards. First, rules are needed to promote the comparative environmental advantages of using natural gas vehicles. For instance, regulations that limit emissions of pollutants from all vehicles can encourage the design of natural gas vehicles that take advantage of the clean-burning characteristics of natural gas. Second, codes requiring manufacturers to use standard materials for construction of natural gas storage cylinders can avoid potential safety problems.

The US Environmental Protection Agency (EPA) and California regulators are engaged in major rule-making processes that take into account the environmental and safety concerns associated with the use of natural gas as a transportation fuel. The US Department of Transportation and the National Fire Protection Association are involved in setting safety standards related to natural gas vehicles, and the Gas Research Institute and the Natural Gas Vehicle Coalition are studying ways to ensure the quality of the natural gas used for transportation.

Emissions Standards: The United States Environmental Protection Agency

The 1990 Clean Air Act Amendments pave the way for implementation of a regulatory regime that will enact environmental and safety controls for natural gas as a transportation fuel. This law directs EPA to establish 45 new sets of regulations addressing every major environmental issue concerning vehicles in general and natural gas vehicles specifically. Most of these rule-making proceedings affect natural gas vehicles only as one subset of all vehicles that must comply with the new regulations; several, however, are particular to natural gas vehicles, including one that requires EPA to promulgate emission standards specifically for natural gas vehicles.

The emission standard issue is connected to the implementation of one of the most significant provisions of the 1990 Clean Air Act Amendments: the restriction of automotive exhaust, or tailpipe emissions, for all cars. Specifically, EPA is required to set new emission standards that will result in a 60 percent reduction in nitrogen oxide pollution and a 39 percent drop in hydrocarbon emissions, compared to vehicles sold in 1992. The new standards are to be phased in starting with 1994 model year vehicles. Individually, both nitrogen oxide and hydrocarbons can cause respiratory problems and burning eyes. In combination, when they interact with sunlight, they form smog which, by reducing visibility and contributing to corrosion of buildings and bridges, is a leading cause of urban environmental blight.

EPA estimates that complying with the stricter emission standards could add $152 to the cost of an average gasoline-powered automobile, due to the addition of pollution control equipment and redesigned engines intended to reduce emissions. By contrast, natural gas vehicles will probably be able to meet the new standards using current natural gas vehicle technology. Thus, the new emission standards are likely to increase the economic competitiveness of natural gas vehicles and promote the use of natural gas as a clean-burning fuel.

However, the new regulations must be drafted so as to clarify the regulatory treatment accorded different classes of hydrocarbon compounds in order for natural gas vehicles to be more competitive. The term hydrocarbon refers to any chemical compound composed primarily of hydrogen and carbon atoms. Gasoline typically is composed of over 150 different hydrocarbons exhibiting a wide range of properties: some are largely inert, while others are toxic or carcinogenic.

Natural gas itself is also composed of hydrocarbons (predominantly methane). Compared to gasoline-powered vehicles, natural gas vehicles emit less of virtually all types of hydrocarbon molecules that cause significant local air pollution problems. Although, with current technology, natural gas vehicle exhaust contains some unburned natural gas, the gas is largely inert in the atmosphere; has not been found to cause health problems at emission levels associated with natural gas vehicle exhausts; and contributes little, if any, to the creation of smog. While total hydrocarbon emissions from natural gas vehicles may exceed those of gasoline vehicles, emissions of the types of hydrocarbons that cause air pollution problems are as much as 85 percent lower in natural gas vehicles.

Prior to the 1990 Clean Air Act Amendments, the hydrocarbon emission standard did not differentiate among various classes of hydrocarbon compounds in vehicle exhausts but, rather, limited total hydrocarbon emissions. The total hydrocarbon standard did not reflect the comparative advantages of burning natural gas. The 1990 Clean Air Act Amendments recognize the difference between natural gas and other more dangerous classes of hydrocarbons. EPA is now free to establish a non-methane hydrocarbon (NMHC) standard for natural gas vehicles that does not include methane when calculating hydrocarbon emissions in vehicle exhaust. For more than two years, EPA developed emission standards for natural gas vehicles and delineated testing procedures to demonstrate compliance. On October 19, 1992, EPA finally issued proposed regulations that include a non-methane hydrocarbon standard appropriate to natural gas vehicle exhaust emissions. The eventual promulgation of these EPA regulations will have an enormous impact on natural gas vehicles, and will boost their widespread commercial viability.

Regulations implementing several other provisions of the 1990 Clean Air Act Amendments could also speed natural gas vehicle commercialization. For example:

- Stricter truck and bus emission standards are contained in the 1990 Amendments. As with regulations implementing the new automobile emission standards, the regulations for these heavy-duty vehicles could favor use of natural gas.
- The 1990 Amendments also require the use of gasoline enriched with oxygen-bearing compounds in 42 cities with high carbon monoxide levels and the sale of lower-polluting "reformulated" gasoline in nine especially polluted cities (see Chapter 1). The higher cost of these special fuels, as compared to conventional gasoline, could further improve the existing price advantage of natural gas as compared to gasoline.
- Natural gas vehicle technology could also become more attractive as a result of new regulations limiting fuel tank evaporative emissions and carbon monoxide emissions during the first few minutes that a car engine is warming up. Evaporation from fuel tanks, which is responsible for up to half of the hydrocarbon emissions from gasoline vehicles, is eliminated with air-tight natural gas storage cylinders. Moreover, significant quantities of air pollutants are emitted during the first couple of minutes of gasoline-powered engine operation. Heat from a warmed-up engine is needed to help vaporize gasoline so that complete combustion of the fuel occurs. Because natural gas is already in a gaseous form when it enters a vehicle engine, the engine warm-up phase does not generate additional pollutants.
- The alternative fuel use mandates for certain vehicle fleets, as outlined in the 1990 Clean Air Act Amendments and the National Energy Policy Act of 1992, are also likely to attract new natural gas vehicle users. Once again, however, this will only happen if regulations are drafted in a manner than will help natural gas vehicles capture a share of the market anticipated under this program.

California Clean Air Standards

Since the first Clean Air Act was passed in 1970, California has consistently been at the forefront of efforts to develop and implement automotive pollution standards that are even stricter than federal ones. Characteristically, while Congress was adopting new emission standards in the 1990 Clean Air Act Amendments (which largely mirrored existing California standards), California was in the process of implementing even more radical new standards of its own.

In September 1990, the California Air Resources Board adopted Low-Emission Vehicle Regulations. The effect of the regulations, which will phase in stricter emission standards over a 15-year period, will, if fully implemented, constitute nothing less than abandonment of the gasoline-powered vehicle in California.

California's new emission standards will be introduced in four stages over the 15-year period.

- The first stage requires that, in model years 1994-1996, at least 10 percent of all vehicles sold in California must be "transitional low-emission vehicles" (TLEVs) that emit 0.125 grams of hydrocarbons per mile (50 percent of the discharges allowed by the Clean Air Act Amendments).
- The second stage requires that "low emission vehicles" (or LEVs) that emit 0.075 grams of hydrocarbons per mile (30 percent of the limit allowed under the Clean Air Act) be phased in by the end of the century.
- The third stage requires phasing in of ultra-low-emission vehicles (ULEVs) that emit 0.04 grams of hydrocarbons per mile (16 percent of the level permitted by the Clean Air Act) by the end of the century. By the year 2003, LEVs and ULEVs must comprise 15 percent of market sales.
- The fourth stage of the standards confirms the truly radical nature of the new California regulations. Starting with 2 percent of new car sales in 1998, "zero emission vehicles" (ZEVs) must be on the streets of California and, by model year 2003, must constitute 10 percent of sales. Only ve-

hicles powered by electricity or, perhaps, by hydrogen energy qualify as ZEVs because their tailpipe emissions are virtually pollutant-free. Because the technology for ZEVs is not completely developed (not a single electric or hydrogen model is in full commercial production), this establishes a tremendous challenge to the automotive industry.

	Permitted Hydrocarbon Discharges	**Percent of Clean Air Act Requirements**
Clean Air Act Amendments	0.25 grams/mile	
TLEVs	0.125 grams/mile	50%
LEVs	0.075 grams/mile	30%
ULEVs	0.04 grams/mile	16%
ZEVs	0 grams/mile	0%

Current gasoline engine technology will probably dominate the TLEV market. Natural gas vehicles, however, may capture a significant portion of the LEV and ULEV categories because the regulation's stricter emission requirements may be impossible to achieve in conventional gasoline-powered engines. Also, because of the similarities they share as users of gaseous fuels, natural gas vehicles could prove to be a stepping stone to hydrogen engine technology and the manufacture of ZEVs. (See Profile 9 for a discussion of the transitional role natural gas can play in the development of hydrogen-based transportation fuels.)

Safety Standards

Natural gas vehicle use also raises other regulatory issues, notably the establishment of proper safety standards for natural gas storage cylinders and construction requirements for the design and operation of natural gas refueling stations.

The excellent safety record for natural gas vehicles notwithstanding (see Chapter 1), obtaining safety certification for new natural gas vehicle parts, especially storage cylinders, is difficult because federal regulations are outdated and unresponsive to the conditions of cylinder use in automotive applications. Storage cylinders are regulated by the US Department of Transportation (DOT). DOT regulations were originally developed for the steel containers generally used to ship compressed natural gas in trucks. DOT has never promulgated regulations specifically concerned with the use of cylinders installed on a vehicle and used to store fuel for use by that vehicle.

DOT regulations currently require frequent retesting of cylinders to inspect for potentially dangerous wear and tear. These tests typically require the costly and time-consuming removal of a natural gas vehicle's cylinders. Alternative testing procedures could provide the same safety checks, without requiring cylinder removal. DOT and manufacturers of natural gas vehicle storage cylinders have been working together for several years to prepare a set of regulations detailing such alternative testing procedures; these regulations are likely to be completed in 1993 or 1994.

A similar situation hampers construction of natural gas refueling stations. Refueling station construction is generally subject to local building codes; these in turn usually require compliance with codes issued by the Massachusetts-based National Fire Protection Association Committee 52 (NFPA-52 codes). First established in 1984 and updated in 1988, NFPA-52 codes again need to be revised in the face of rapidly changing natural gas vehicle technology. Proposed revisions were developed by natural gas vehicle advocates, spearheaded by member companies of the Natural Gas Vehicle Coalition in 1991, and are being reviewed by the NFPA. Adoption in 1993 is likely, thereby easing one obstacle to commercialization of natural gas vehicle technology.

Gas Composition

A final area of concern, more in the future than today, involves regulations to ensure the quality of natural gas used in natural gas vehicles. All natural gas is not the same. Methane is the principal component of natural gas. Mixed in with the methane are trace amounts of water, inert molecules, and a few other hydrocarbons such as butane and propane that normally are mixed with natural gas when it is produced. Moreover, some utilities deliberately add propane to natural gas pipelines during periods of high demand in order to increase gas supplies, a procedure called "propane peak shaving." A 1991 survey of natural gas composition in 26 United States cities, performed by the American Gas Association Laboratories, found that non-methane molecules varied from 3 percent to 17 percent of the total weight of natural gas.

Varying and inconsistent natural gas composition has not yet created problems for natural gas vehicle users, but it could in the future. As natural gas vehicle engine technology becomes more refined to optimize the advantages of natural gas as a vehicle fuel, the engines may become more finicky about the fuel they burn. The result could be increased engine knock or other performance problems when natural gas containing higher levels of non-methane components is encountered. Furthermore, when the concentration of heavier molecules in natural gas, including water and propane, approaches 15 percent, freezing during compression and decompression and clogging of fuel lines can occur. Water can also corrode cylinder walls, especially in high-grade steel cylinders, potentially creating safety concerns through long-term cylinder usage. Improving gas quality with dehydration technology and selective filtering of non-methane hydrocarbons is technically simple, but it adds a costly extra step to natural gas processing.

Currently, natural gas quality is regulated by standards, including NFPA-52 codes, to prevent clogging under conditions found in natural gas pipelines. Water, for example, is limited to 7 pounds per million cubic feet of natural gas. As natural gas vehicle technology develops, a review and revision of gas quality standards may be necessary, at least with regard to natural gas destined for the automotive market. On February 13, 1992, the Gas Research Institute convened a workshop to further explore this issue, and a technical committee has been formed within the Natural Gas Vehicle Coalition to develop new standards if they prove necessary.

Profile 21 Removing Local Natural Gas Vehicle Use Restrictions: New York City

Individual cities and other municipalities are free to establish their own laws and regulations as long as they do not conflict with overriding federal and state rules. There are myriad sets of land use and zoning laws, building codes, and municipal ordinances that affect — and in many instances discourage — natural gas vehicle use. Thus, one productive route for encouraging natural gas vehicle development and use is to revise the laws and regulations affecting them at the local level.

Natural gas vehicle operators often encounter exacting and sometimes perplexing local requirements that limit and in some cases prevent natural gas vehicle use. For example, some communities prohibit construction of natural gas refueling stations, even in areas where gasoline refueling stations are permitted, while others have enacted overly protective safety precautions for natural gas vehicles (relative to the precautions applied to other energy activities with similar risks), such as building restrictions near natural gas vehicle refueling compressors or bans on parking natural gas vehicles in enclosed garages.

Many such local regulations, like the sale-for-resale prohibitions discussed in Profile 14, were promulgated decades ago and were not drafted with natural gas vehicles in mind. Historically, for example, safety precautions for natural gas storage were aimed at regulating large energy storage depots, not the smaller quantities of natural gas taken from local distribution lines and compressed for sale at vehicle refueling stations. Similarly, local vehicle use restrictions were aimed at large, natural gas-carrying tanker trucks, and not at passenger vehicles equipped with a few comparatively small storage cylinders containing their own fuel supply.

A major task if natural gas vehicles are to gain widespread public acceptance is the identification, examination, and revision of local regulations that impede natural gas vehicle use. All too many natural gas vehicle operators have first learned about a local rule when a building or fire inspector knocked on the door with a citation. To avoid this situation, it is necessary to identify and examine possible local restrictions. Thus, a well-designed fleet conversion program or a marketing strategy aimed at converting individual vehicles to natural gas use would first include a thorough review of local codes and regulations. Additionally, calls to fleet operators or equipment vendors in other areas could reveal stumbling blocks encountered elsewhere.

Removing local obstacles to natural gas vehicle use is a two-step process. First, the true risks posed by natural gas vehicle use must be analyzed. Such an analysis should take into account the comparative risks of conventional vehicle use that are not similarly restricted by local regulations. It can also be helpful to assess natural gas vehicle risk by comparison to the risks of the activity which rule makers had in mind when a regulation was first enacted. Second, local political support for natural gas vehicles must be established to ensure that existing rules are revoked and replaced with legislation encouraging natural gas vehicle use without compromising safety or other legitimate public concerns. Establishing political support involves public education and political lobbying.

Bridge and Tunnel Restrictions: New York City

A New York City controversy affecting natural gas vehicle use on certain roadways offers a textbook example of how well-intentioned city leaders can unintentionally create a regulatory compliance problem for natural gas vehicle owners and operators.

The controversy dates back to a May 1949 collision and fire in the Holland Tunnel connecting New York and New Jersey. The accident involved ten heavy-duty freight trucks. One of the vehicles was a tanker truck loaded with highly flammable carbon disulfide; it ruptured upon collision and exploded into flames that quickly engulfed the other trucks. The accident closed the tunnel for one week, and over 650 tons of charred debris were eventually removed, including the truck bodies which, in the intense heat, had fused together into a solid metal mass. Amazingly, there were no deaths.

In the aftermath of the accident, a local regulation was adopted by the Port Authority of New York and New Jersey, the agency governing use of the bridges and tunnels connecting the two states. The rule prohibited travel by vehicles containing a wide range of hazardous materials, including compressed natural gas, through Port Authority tunnels and on the lower level of the George Washington Bridge. These regulations were expanded to include other tunnels and bridges within New York City. Because four of New York City's boroughs (including Manhattan) are islands or are on islands, these restrictions seriously complicated the transportation of hazardous materials between New York City and neighboring New Jersey and Long Island.

When alternative fuel vehicles were first tested in New York City decades later, the Port Authority ruled that even natural gas filled into small, onboard natural gas storage cylinders qualified as a hazardous material under the regulation. Thus, natural gas vehicles were prohibited from using many of the city's critical bridge and tunnel thoroughfares. Although demonstration programs continued in the late 1980s using routes that avoided these roadways, the travel restriction loomed as an obvious barrier to widespread natural gas vehicle use in New York City.

Attempts to convince the Port Authority to change its hazardous materials transport rules failed. Hence, proponents of natural gas vehicles decided to conduct an analysis that would document the actual risks posed by natural gas vehicle use. In 1989, the New York State Energy Research and Development Authority joined Brooklyn Union Gas Company and Consolidated Edison Company in funding this risk assessment, which was conducted by Ebasco Services, Inc. It focused on tunnels, since they were acknowledged to pose a greater accident risk than bridges for fuel-related accidents.

The findings of the risk assessment were published in 1990 as *Safety Analysis of Natural Gas Vehicles Transiting Highway Tunnels*. The study analyzed the potential for damage from collisions of gasoline-fueled vehicles and natural gas vehicles under a variety of tunnel conditions. Its assessment showed that natural gas would be withdrawn from the scene of an accident (through a tunnel's ventilation system) faster than gasoline. Moreover, the assessment noted that specific combustion characteristics of

New York City Tunnels and Bridges

Transportation regulations, now lifted, prohibited the use of natural gas vehicles in four tunnels and on the lower levels of two bridges in New York City:

- George Washington Bridge (lower level)
- Lincoln Tunnel
- Queens Midtown Tunnel
- Holland Tunnel
- Brooklyn-Battery Tunnel
- Verrazano Narrows Bridge (lower level)

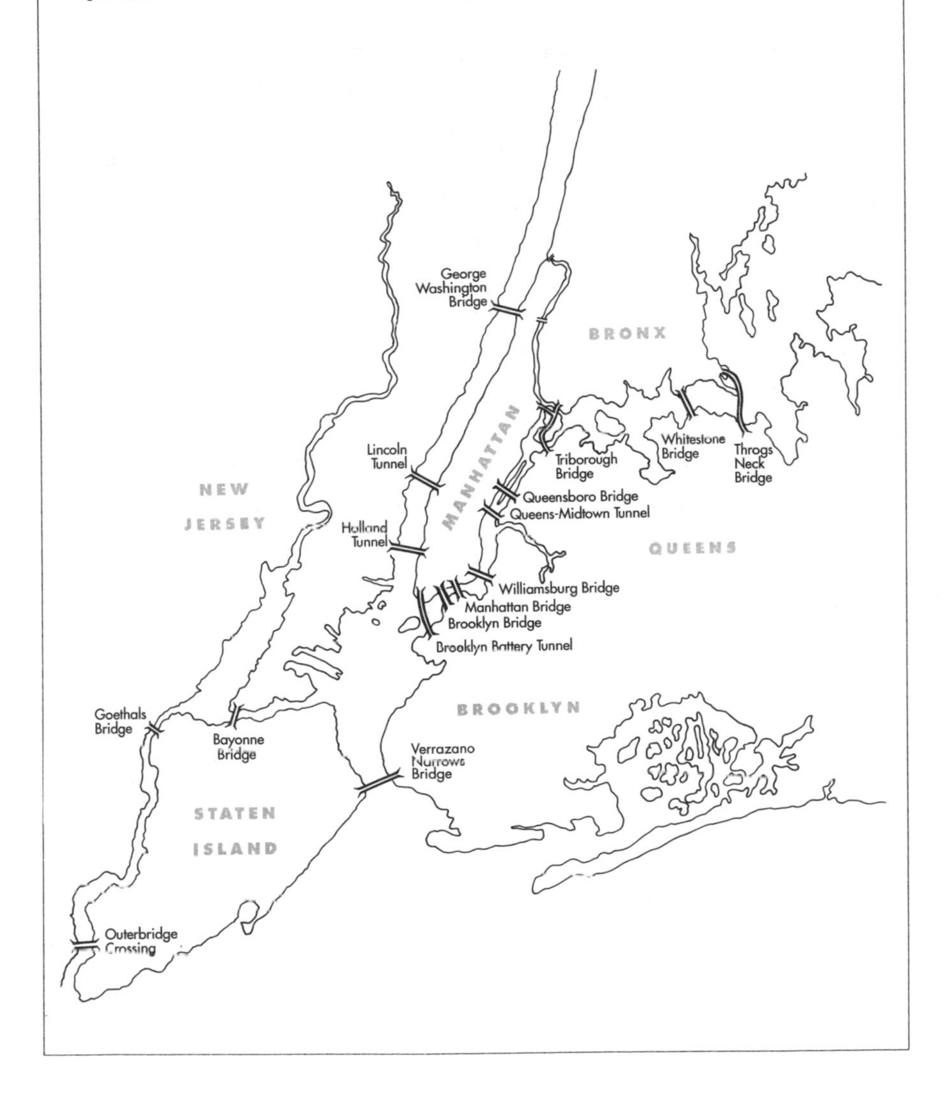

natural gas, such as its high flammability threshold and its high ignition temperature compared to gasoline, provided additional safety benefits over gasoline. The report concluded that "data examined in this study indicated that the fire, fire-related injury and fire-related fatality rates of CNG [compressed natural gas] vehicles are substantially less than gasoline vehicles. . . . This shows that the overall risk of a CNG vehicle in a tunnel is comparable to or less than a gasoline vehicle depending on the hazard category."

The safety advantages of natural gas were most readily apparent in the most extreme hazard category examined in the study. An extremely hazardous accident involving a gasoline-powered vehicle is one resulting in a so-called boiling liquid expanding vapor explosion (BLEVE). An equivalent accident involving a natural gas vehicle is a fuel line rupture resulting in the rapid discharge and ignition of natural gas under pressure. When Ebasco compared the possible damage from these two worst-case accidents, it found that the gasoline BLEVE "can result in a thermal release rate 10-100 times greater than CNG ignition. The BLEVE power release from a small [gasoline-powered] sedan exceeds the worst case scenario for the CNG bus."

Armed with the Ebasco study, New York City natural gas vehicle advocates again lobbied for variances from or revisions to the tunnel and bridge restrictions affecting natural gas vehicles. This time their efforts were successful. Effective March 26, 1990, the Port Authority revised its regulations. A few months later, the Triborough Bridge and Tunnel Authority, which has jurisdiction over some of the bridges and tunnels within New York City, also lifted its restrictions. Since then, natural gas vehicles have been free to traverse the bridges and tunnels linking the two states and the boroughs, and a natural gas-powered future is more likely than before in New York City.

Rooftop Fuel Storage

Another example of restrictive regulation, this one inhibiting development of natural gas bus technology, illustrates the importance of examining the original intention of regulations. Natural gas buses with a driving range comparable to that of gasoline or diesel buses require up to ten natural gas storage cylinders. Although some cylinders can be attached under the chassis of the bus, engineers have designed an alternative involving installation of the tanks on the roof of the bus. The tanks are encased in a compartment built for this purpose, hiding them from view. Prototype buses with this design were built in 1990 by Bus Industries of America.

The problem is a regulatory requirement in some United States locations prohibiting fuel storage anywhere except under a vehicle. This regulation was established with liquid fuels in mind. Under-vehicle fuel storage eliminates the possibility of a leak of liquid fuel dripping into the passenger compartment, where a fuel fire would have deadly consequences. While gasoline and diesel fuel obviously drip down in response to gravity, a leak of natural gas disperses upward because it is lighter than air. Thus, it appears that rooftop fuel storage for natural gas would be safer than storage on the undercarriage, by the same logic that rooftop storage of liquid fuels should be avoided.

Logic aside, rooftop fuel storage still faces legal restrictions in the United States. These are gradually being overturned through the work of natural gas vehicle advocates. The first demonstration of natural gas buses with rooftop fuel storage, however, took place in Toronto, Canada, where there is no ban on this design.

Profile 22 Investigating Unanswered Environmental Issues: The International Energy Agency

The environmental edge natural gas enjoys compared to gasoline or diesel fuel is a major reason for its appeal as a transportation fuel. Specifically, it is the desire for cleaner air, more than any other reason, that recommends natural gas vehicle technology to the public. As detailed in INFORM's 1989 study *Drive For Clean Air* and discussed in Chapter 1, tests of emissions from automotive tailpipes confirm that discharges of most major vehicular air pollutants are greatly reduced in natural gas vehicles. Natural gas vehicles produce negligible amounts of evaporative emissions, which account for about half of all air pollution from gasoline-powered vehicles. In addition, the problems of tanker crashes, oil spills, and leakage under storage tanks obviously do not apply to natural gas vehicle technology.

If natural gas vehicles are to gain widespread public acceptance and use in the long term, it is therefore necessary that the environmental consequences of natural gas use be completely understood. In particular, natural gas vehicles cannot secure a central place in the United States transportation future unless unanswered environmental issues are investigated, assuring the public that natural gas use and natural gas vehicle technology do not contain some fatal — but as yet undetected — flaw.

The discovery of previously undetected flaws is not uncommon in emerging energy industries. What may appear to be a very promising technology, relatively free of environmental problems, can be stopped dead in the face of unanticipated public concern. In the case of methanol-powered vehicles, for instance, nagging concerns about high emissions of formaldehyde, a suspected human carcinogen, continue to raise questions about the environmental suitability of using methanol as a fuel. This is true despite the fact that there is considerable evidence that total air pollution caused by methanol vehicles is less than that caused by vehicles powered by gasoline and diesel fuels.

Environmental scrutiny of natural gas vehicles has historically focused on natural gas vehicle tailpipe emissions of the major air pollutants regulated under the Clean Air Act, such as hydrocarbons, carbon monoxide, and nitrogen oxides. However, natural gas vehicles emit other pollutants, unregulated as of mid-1992, of potential environmental concern, most notably emissions of the so-called greenhouse gases implicated in global climate change.

In addition, some discharges of regulated and unregulated pollutants occur during the production and distribution of natural gas destined for use in natural gas vehicles, adding to the total environmental impact of natural gas during its entire fuel cycle from production through end use. Carbon dioxide is thought to be the principal cause of global warming, accounting for as much as half of all greenhouse gas emissions. While natural gas vehicles emit about 15 percent less carbon dioxide than conventional gasoline vehicles, their discharges are still significant.

Natural gas vehicles also emit small quantities of another pollutant of potential concern in regard to global climate change: methane. Although not regulated as a pollutant under the Clean Air Act, methane, which is essentially unburned natural gas, is thought to account for about 18 percent of all total global warming. Like carbon dioxide, methane absorbs infrared radiation in the stratosphere, retaining heat that would otherwise dissipate into outer space. There is much less methane than carbon dioxide in the atmosphere (about 1 molecule per million molecules in the atmosphere is a methane molecule, as opposed to 345 molecules per million for carbon dioxide). However, each methane molecule is thought to absorb up to 80 times more heat than a molecule of carbon dioxide. On the other hand, methane is much less durable, remaining in the atmosphere only a decade or two before decomposing, while carbon dioxide degrades slowly over the course of a century. Combining the impact of its higher absorbance and lower durability, methane appears to be about 10 to 20 times more potent as a greenhouse gas than carbon dioxide on a molecule by molecule basis.

Methane primarily comes from non-energy-related, natural sources. Agricultural operations, the cattle industry, the digestion of wood by termites, and other natural processes of decomposition may account for as much as 75 percent of global methane emissions. Much of the remaining 25 percent comes from methane inadvertently released during coal production and from the flaring, or burning off, of natural gas during oil production. Although United States natural gas production and use is currently a small contributor to global methane emissions, a major transition to natural gas as a transportation fuel could lead to an increase in methane emissions, which could further increase global climate change.

Greenhouse Gas Emissions: International Energy Agency Study

In order to clearly understand the comparative implications for global climate change of natural gas vehicle use (or for any vehicle type, for that matter), it is useful to compare sources of carbon dioxide and methane emissions at each stage in the production and use of natural gas and gasoline. An emerging branch of environmental science is attempting to do just that by identifying and characterizing the environmental impact of these gases at each stage of the cycle of energy production and use. Called fuel-cycle environmental analyses, these investigations provide a complete picture of the total environmental repercussions of energy production and use.

One of the most ambitious projects is a multiyear study underway at the International Energy Agency (IEA) in Paris. Tentatively titled, as of October 1992, *Transport System Responses in the OECD: Greenhouse Gas Emissions and Road Transport Technology*, this study estimates and compares greenhouse gas emissions caused by the production and use of conventional automotive fuels and a variety of alternative transportation fuels. The study, scheduled for publication late in 1993, is drawn in part from another 1992 study (published by the Argonne National Laboratory) entitled *Emissions of Greenhouse Gases from the Use of Transportation Fuels and Electricity*.

Using a natural gas vehicle market scenario for the year 2005 developed by IEA, the analysis looked at the entire fuel cycle, including vehicle emissions, fuel production and distribution, and production of the materials utilized in the fuel cycle (including, for example, air pollution emitted during the construction of an oil tanker or a natural gas pipeline). When all these factors are taken into account, the study concluded that natural gas vehicles will contribute 14.4 percent less to global warming than gasoline-

Comparison of Greenhouse Gas Emissions throughout the Fuel Cycle: Estimate, North American Market, 2005

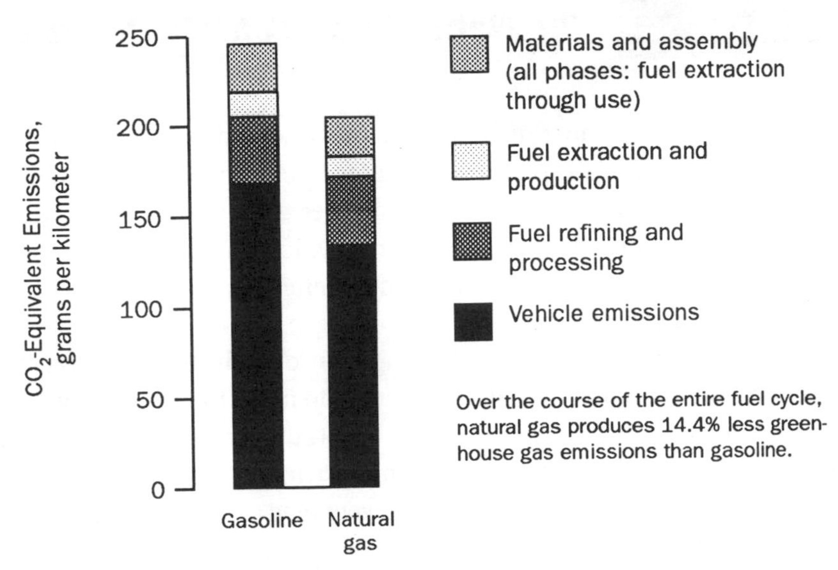

Source: International Energy Agency

powered vehicles, even though tailpipe emissions of methane from the average natural gas vehicle will be 25 times higher than those from the average gasoline-powered vehicle. Tailpipe emissions of carbon dioxide will continue to be 15 percent lower from natural gas vehicles than from gasoline-powered vehicles. Thus, the IEA study suggests that methane emissions from natural gas vehicle tailpipes need not hinder the development and expanded use of natural gas vehicles relative to gasoline vehicles.

Scientific questions about multiple matters relating to methane remain: for example, the exact potency of methane as a greenhouse gas, the chemistry of methane in the upper atmosphere, and the actual sources and extent of global methane emissions worldwide. More information needs to be collected and analyzed. The question of methane emissions from natural gas vehicles should be addressed forthrightly and should remain a top priority in the fight to improve air quality.

Profile 23 Providing a Knowledgeable Voice for Natural Gas Vehicle Interests: The Natural Gas Vehicle Coalition

Advocates of all manner of United States social and political issues have found that effectiveness partly requires focused, forceful representation of their views to a variety of different audiences. Unless such advocates can speak intelligently and clearly with a single voice, they run a very great risk of being trampled beneath the loud and forceful chants of better organized adversaries or competitors.

Profiles 20 and 21 discussed the importance of changing federal, state, and local laws, regulations, and ordinances to make them more conducive to natural gas vehicle use. If this is to occur, those with an interest in the development and use of natural gas vehicles (from natural gas and automotive companies and natural gas vehicle owners and operators to environmental advocates and municipal planners seeking cleaner and less costly fuels) must organize to represent the case for natural gas vehicles in a clear and intelligent way to elected officials, regulators, and the general public. A lobby for natural gas vehicles has been slow to develop, but a unified voice advocating the value of natural gas vehicles is now emerging.

Interest in natural gas vehicles comes from a disparate group of enterprises with little or no experience working together. In part, natural gas vehicle supporters must unite with advocates of other alternative transportation fuels. A unified interest group for natural gas vehicles would include the natural gas utility industry (the major economic beneficiary of natural gas vehicle technology), the automotive industry, environmentalists, consumers, and the many small businesses now refining various kinds of natural gas vehicle technology.

The automotive industry has historically been slow to show its support for natural gas vehicles, as discussed in the introduction to Chapter 5. Not only has the automotive industry been slow to change its accustomed practices but, because natural gas was not used as an automotive fuel until very recently, the automotive industry has had little history of working with natural gas utilities for common goals. The natural gas industry itself was slow to become an advocate of natural gas use for transportation purposes. Moreover, environmental and consumer groups have often locked horns with both the natural gas utility and the automotive industries over issues such as pollution control, safety, and cost. An added problem, finally, is that the emerging small businesses that comprise much of the natural gas vehicle and natural gas refueling equipment industries often have neither spare time nor resources to participate actively as members of organizations representing natural gas vehicle interests.

For most of the 1980s, a committee of the American Gas Association (AGA) provided the focus for natural gas vehicle interests nationwide. AGA sponsored an annual natural gas vehicle conference, starting in 1983, at which the Natural Gas Vehicle Committee played a limited role in lobbying for legislation and regulations affecting natural gas vehicle technology; it also provided some technical and educational services. Participation on the committee was limited to AGA member companies, not all of which were uniformly avid supporters of natural gas vehicles.

The Natural Gas Vehicle Coalition

In 1988, the first organization to solely promote the use of natural gas as a transportation fuel was formed. The Natural Gas Vehicle Coalition (NGVC) began as a direct offshoot of AGA's Natural Gas Vehicle Committee, and its roots are firmly in the natural gas utility industry: its 20 founding members were all natural gas utility companies and, as of October 1992, NGVC headquarters were located at an office building occupied by the AGA in Arlington, Virginia.

However, NGVC's mission is more focused than that of the AGA committee, and it aims to unite the disparate entities with an interest in promoting natural gas vehicle technology. By October 1992, NGVC membership had grown to about 150, including 81 natural gas vehicle equipment suppliers and 41 natural gas utilities. Other categories of membership include natural gas pipeline companies and associate members such as government agencies, trade organizations, nonprofit groups, and individuals. Three state governments (New Mexico, Kansas, and Texas) send representatives to NGVC. Trade association members include the International Association for Natural Gas Vehicles, the Petroleum Equipment Institute, the Australian Gas Association, the Gas Appliance Manufacturers Association, and the Steel Tank Institute. Two schools also belong to NGVC: Hocking Technical Institute (in Ohio) and the University of Oklahoma Energy Center. No environmental organizations were official members as of October 1992. Annual membership fees range from $500 for nonprofit and government organizations to $20,000 for large natural gas utilities.

The specific objectives of NGVC are to stimulate the natural gas vehicle market through strategic marketing; to influence and support government policies favorable to natural gas vehicles; and to coordinate research, development, and demonstration of natural gas vehicle technologies. NGVC's day-to-day affairs are conducted by an executive director assisted by a four-person staff. Overall direction for NGVC is provided by a board of directors assisted by the Executive, Managing, and Industry Advisory Committees. In addition, four committees, staffed by NGVC members, have been established: Government Affairs, Technology, Market Development, and Public Relations and Communications.

NGVC's Government Affairs Committee has spearheaded much of the rapid evolution of United States energy policies and programs regarding alternative transportation fuels. Its Federal Affairs Subcommittee lobbied vigorously for the inclusion of provisions relating to alternative transportation fuels in the 1990 Clean Air Act Amendments. In 1991 and 1992, its lobbying efforts focused on the National Energy Policy Act and on participating in the wide range of regulation-setting processes affecting natural gas vehicle use underway at federal agencies. Its State and Local Affairs Subcommittee provides information and assistance to state governments interested in establishing alternative transportation fuel programs.

NGVC's Technology Committee has far-reaching responsibilities affecting all aspects of natural gas vehicle hardware, including natural gas vehicle conversion equipment, vehicles manufactured as dedicated natural gas vehicles, refueling stations, and natural gas distribution and storage systems. The development of codes and standards for natural gas vehicle technology, especially engine emission standards and storage cylinder safety specifications, is addressed by the Technology Committee. The Technology Committee is also working closely with the major automotive manufacturers to develop assembly line-produced vehicles. In 1991, Ford and General Motors became the first two automobile manufacturers to join NGVC. (See Profile 2 for a discussion of natural gas vehicle assembly line production.)

Increased commercialization of natural gas vehicle technology in private and government fleets is the primary responsibility of NGVC's Market Development Committee and its two subcommittees, Market Services and Infrastructure Development. The committee provides market analyses and information to prospective users of natural gas vehicle technology and also conducts national market analyses and economic assessments to help identify potential natural gas vehicle users.

Finally, NGVC's Public Relations and Communications Committee develops press materials, including information packets and press releases. It also organizes events and programs to increase public awareness of natural gas vehicle technology and projects underway involving natural gas vehicle use. The work of this committee is discussed in more detail in Profile 24.

During its short existence, NGVC has provided a vital service to all constituencies with an interest in the eventual widespread use of natural gas vehicles. Gradually, it is likely that several additional organizations will emerge, representing specialized natural gas vehicle interests in both the private and public sector.

Other Advocacy Groups

Other knowledgeable advocacy groups have already begun to appear. On January 22, 1991, the California Natural Gas Vehicle Coalition was founded by that state's three largest natural gas utility companies. Headquartered in Sacramento, the California NGVC concentrates solely on issues relating to natural gas vehicle use in California; by October 1992, it had produced a California Natural Gas Vehicle Strategic Commercialization Plan.

Similarly, knowledgeable advocacy groups appear to be forming in the public sector. In early 1992, the governors of several southwestern states, including Texas, Oklahoma, and New Mexico, joined to promote the establishment of a Natural Gas Vehicle Zone. The zone would consist of refueling stations strategically situated along the region's interstate highway corridors. (See Profile 12 for more information about the Natural Gas Vehicle Zone.)

In April 1992, a first-of-a-kind collaborative project involving natural gas utilities and environmental organizations culminated with the publication of a joint study, *Alternative Energy Future*. The participants included the American Gas Association, the Alliance to Save Energy, and the Solar Energy Industries Association. The study examined a variety of future energy policies for the United States. It concluded that expanded use of energy-efficient and renewable energy technologies, coupled with a 40 percent increase in natural gas usage, could stabilize United States energy consumption at 1990 levels, while reducing pollution and stimulating a 50 percent growth in national economic activity. This united effort could mark the beginning of increased collaborative activities between natural gas interests and environmental constituencies.

Profile 24 Promoting Public and Professional Education about Natural Gas Vehicles: Specialized Conferences and Publications

One barrier to the development and commercialization of natural gas vehicles that can be easily overcome is the barrier of insufficient public, professional, and media awareness and understanding. The general public is often uninformed of the importance of and need for the use of alternative fuel technology and the suitability of natural gas as an heir to gasoline as a transportation fuel. The vast majority of the driving public does not know that natural gas vehicles work well, are environmentally clean, operate safely, and use a fuel that is one-third cheaper than gasoline. Moreover, the concept of how a gaseous fuel can replace a liquid fuel is still a mystery to most people, and the fear that compressed gas storage must pose greater dangers than gasoline storage is a widely held but unfounded public perception. In short, until the considerable advantages of natural gas become widely known, natural gas vehicles are not likely to be in great demand.

The problem is not limited to the general public. Energy policymakers in government and managers who make decisions about fleet vehicle use often lack the information they need to make informed choices about which alternative fuels to use and how best to design effective alternative fuel vehicle programs. Indeed, because alternative fuel technology is progressing so rapidly, natural gas vehicle specialists are sometimes hard-pressed to remain informed about developments in the field.

Specialized Conferences

The increasing number of specialized conferences and seminars demonstrates the thirst for natural gas vehicle information and education: professional natural gas vehicle education is increasing at all levels.

For instance, attendance at annual natural gas vehicle conferences, which have been sponsored for a decade by the American Gas Association (AGA), increased fivefold from 1987 to 1988 alone. The 1987 Indianapolis meeting was attended by 50 people, mostly technical experts. In 1988, the same event drew 250 participants to Indianapolis. The meetings were then moved to attract even larger audiences. For two years, AGA's natural gas vehicle conference was held in Washington, DC and, in 1991, in San Francisco. The 1992 meeting was held outside Disney World in Orlando, Florida, and attracted 700 attendees.

In addition to national AGA conferences, a dozen or more major alternative transportation fuels conferences are now held annually in the United States. For example, the third annual Texas Alternative Vehicle Fuels Market Fair, held in Austin in April 1992, was attended by about 1600 people — probably the largest audience ever to attend a conference about any alternative transportation fuel. Outside the United States, in June 1992, the Canadian-sponsored Windsor Workshop on Alternative Fuels held its ninth annual technical conference in Toronto. The International Association for Natural Gas Vehicles convened its third World Natural Gas Vehicle Conference in Goteborg, Sweden in September 1992, with an attendance of about 450 people. The first two international conferences were held in Sydney, Australia and Buenos Aires, Argentina, respectively. In October 1992, the first British conference on natural gas vehicles was convened in London.

In addition, the volume of available information about natural gas vehicles in specialized, technical publications and journals is rapidly growing. Similarly, every year more demonstrations are held to showcase the advantages of natural gas vehicles and increase their visibility. As professional awareness and understanding of natural gas vehicles grow, visibility spreads to the more general media and the general public.

Specialized Publications

By now, virtually every major newspaper and magazine that covers energy or environmental issues has devoted at least one article to alternative transportation fuels. Nonetheless, although reporting in the press is becoming more frequent, it is still irregular, even in the specialized energy trade press. Since 1989, several publications have been launched that address the need for more comprehensive information about alternative transportation fuels. Some of them are described here:

- *Clean Fuels Report* is a quarterly journal started in 1989 and published in Niwot, Colorado. Each edition devotes nearly 200 pages to in-depth reporting of all aspects of alternative transportation fuels.
- *Natural Gas Vehicle* is a bimonthly magazine published by the American Gas Association in Arlington, Virginia. It is devoted almost solely to reporting events of significance to the natural gas vehicle industry, such as the opening of new refueling stations and the passage of state laws encouraging use of alternative transportation fuels. *NGV News*, published monthly, is another AGA publication devoted to natural gas vehicles.
- *Natural Gas Fuels* is a monthly magazine published in Denver, Colorado since August 1992. It includes feature articles about all aspects of natural gas vehicle use, as well as shorter news and events columns.
- *New Fuels Report* is a weekly newsletter produced in Washington, DC and *Environmental Vehicles Review* is a monthly newsletter published in Oakland, California. Both publications were founded in 1990 and are dedicated to covering events in the alternative transportation fuels field.
- *Clean Fuel Vehicle Week*, published in San Clemente, California, and *CNG Monthly*, published in Austin, Texas, are other new publications.

All of these publications are fairly technical and are aimed largely at business and government audiences with an existing interest in natural gas vehicles.

Natural gas powered the camera trucks leading the runners at the Pittsburgh Marathon on May 3, 1992. Photos: Consolidated Natural Gas

Public Visibility

Special media events are increasingly being effectively used to increase the public visibility of natural gas vehicles as a viable transportation alternative. In 1991, for example, the lead car in the Boston Marathon was a natural gas vehicle; in 1992, a natural gas vehicle led the way in the Pittsburgh Marathon. A racing car powered by natural gas, called the "Natural Gasser," has been run on test tracks at the Indianapolis Speedway and the Bonneville Salt Flats in Utah. Plans are underway to make natural gas vehicles the official car of the 1996 Summer Olympics in Atlanta.

In June 1991, the Natural Gas Vehicle Challenge, a road rally held in Oklahoma City, brought together members of the general public, university students, and technical experts for a highly publicized natural gas vehicle event. Students from 24 engineering universities competed in the rally. Each group of students modified a conventional Sierra pickup truck into a natural gas vehicle; the pickups were donated by General Motors Corporation. During the two-day rally, the vehicles were tested to measure a variety of performance characteristics, including acceleration, emissions, fuel economy, and road-handling ability. Awards were given in about 10 categories at a videotaped press conference and ceremony that concluded the event. The rally was staged to coincide with the First Continental Congress on Natural Gas Vehicles, held simultaneously at the University of Oklahoma.

A second Natural Gas Vehicle Challenge was held in May 1992. Twenty-one colleges and universities participated in the event, staged in Michigan and Ontario, Canada. The third Challenge is scheduled for June 1993 in Austin, Texas.

In another road rally in Australia in February 1992, a natural gas-burning Honda set the world distance record for a natural gas vehicle. The car traveled 400 miles on natural gas from just one storage cylinder containing approximately five equivalent gallons of gasoline. The car averaged 80 miles per equivalent gallon over a course that included climbing Mount Kosciusko, Australia's highest peak.

Profile 25 Supporting International Information Exchange and Cooperation: The International Association for Natural Gas Vehicles

Natural gas vehicle technology addresses energy and environmental issues that are global in nature. Dependence on oil as a transportation fuel is a worldwide concern, with most of the world getting oil from a very small region. Furthermore, the use of oil contributes to the greenhouse warming of our planet, another issue with global impact.

Hence, there is an international movement to diversify fuel sources; many countries are undertaking alternative fuel programs, and governments and businesses around the world are developing technologies for natural gas vehicles. For example, FuelMaker, the vehicle refueling appliance, was developed in Switzerland, is manufactured in Canada, and is now being marketed in the United States. (See Profile 13 for a more complete discussion of refueling appliances.) Italian-built natural gas vehicle conversion systems are the largest sellers in New Zealand. In Scandinavia, the Co-Nordic Natural Gas Bus Project, sponsored by several Scandinavian countries, has as its goal the development of natural gas bus technology that meets the strict standards set in the United States 1990 Clean Air Act Amendments. With natural gas vehicle programs springing up around the world, international information exchange and cooperation are essential. Without them, development of effective technologies will surely be inefficient.

The International Association for Natural Gas Vehicles

The International Association for Natural Gas Vehicles (IANGV) is the first and, as of October 1992, the only organization that exists to serve global natural gas vehicle interests. Founded in 1986, IANGV now has 180 members from 30 countries. New Zealand, which thanks to government leadership and a large effort had the world's most progressive natural gas vehicle program in the 1980s, provided the initial impetus for the establishment of IANGV. The IANGV Secretariat is located in that country's largest city, Auckland.

IANGV is governed by an International Executive Council that includes one delegate from each of the countries represented in the organization's membership. The total IANGV membership from each country constitutes a separate National Council. Each National Council elects its delegate to the International Executive Council which, in turn, selects the Secretary General, who is responsible for handling the daily tasks of running IANGV. Since its creation, Garth Harris, a long-term leader in New Zealand's natural gas vehicle program, has served as Secretary General. IANGV has nine membership categories, and dues range from $60 to $1200 per year.

With an annual budget of $100,000 and a paid staff of only two employees, IANGV is still a small organization. However, its mission is broad and its accomplishments to date are impressive. The objective of IANGV is to increase natural gas consumption by the world's transportation systems. To this end, the organization conducts programs that highlight the efficiency, safety, durability, and environmental and economic advantages of natural gas vehicles. Its 1992 program included five specific activities:

- Collection and dissemination of information;
- Sponsorship of conferences and symposia;
- Promotion of uniform natural gas vehicle standards and codes;
- Encouragement of natural gas vehicle research and commercialization by original equipment manufacturers in the automotive industry; and
- Identification and setting of priorities for critical natural gas vehicle research and development needs.

During its first six years of existence, the first three program areas have dominated the work of IANGV. Its showcase activity has been organizing and conducting three international natural gas vehicle conferences. The first World Natural Gas Vehicle Conference was held in Sydney, Australia in October 1988; about 250 people participated. The three-day event included an exhibition of natural gas vehicle technology by more than 30 companies.

The second World Natural Gas Vehicle Conference, in October 1990, attracted more than 300 people from 34 countries to Buenos Aires. More than 60 professional papers were presented at the conference and published in a three-volume set of conference proceedings. To mark the significance of the event, the mayor of Buenos Aires used the forum of the opening ceremonies to announce a local ban on the sale of diesel-powered buses and a requirement that new taxis operate on natural gas within the city limits. The third World Natural Gas Vehicle Conference, held in Goteborg, Sweden in September 1992, drew about 450 people.

In addition to sponsoring conferences, IANGV publishes a quarterly newsletter and an annual yearbook as part of its education and information exchange efforts. In 1990, IANGV published an overview of natural gas vehicle technology around the world entitled *Natural Gas Vehicles 1990;* it periodically publishes other reports. In October 1991, for example, it released the results of a survey and inventory of natural gas vehicle programs worldwide. The inventory identified six countries as each having more than 20,000 operating natural gas vehicles. They include Italy (235,000 natural gas vehicles), Argentina (100,000), New Zealand (50,000), the United States (30,000), and Canada (26,000). In addition, more than 200,000 natural gas vehicles are thought to be operating in the former states of the Soviet Union, but an accurate inventory was not obtainable for the survey.

At present, IANGV's other major program is a cooperative effort to develop standardized equipment specifications for critical natural gas vehicle and refueling station components. For instance, a Cylinder Task Force was set up in 1990 to coordinate work related to natural gas storage tank specifications. In 1991, the work of this group led to the establishment of a broader-based Technical Committee within IANGV. In addition to helping to coordinate cylinder standards, the Technical Committee deals with other issues, including natural gas vehicle conversion equipment certification, standardization of refueling nozzles, and establishment of natural gas quality specifications. This program also aims to establish internationally recognized product safety, performance, and testing codes.

Bibliography

American Gas Association. "An Analysis of the Economic and Environmental Effects of Natural Gas as an Alternative Fuel." Arlington, Virginia: December 15, 1989.

American Gas Association. *1991 Gas Facts*. Arlington, Virginia: 1992.

California Council for Environmental and Economic Balance. *Alternative Fuels as an Air Quality Improvement Strategy*. Sacramento, California: November 1987.

California Energy Commission. *AB 234 Report: Cost and Availability of Low-Emission Motor Vehicle Fuels*. Sacramento, California: August 1989. *AB 234 Report Update*. August 1991.

DeLuchi, Mark. *Emissions of Greenhouse Gases from the Use of Transportation Fuels and Electricity*. Argonne, Illinois: Argonne National Laboratories, 1992.

DeLuchi, Mark, *et al*. *Methanol vs. Natural Gas Vehicles: A Comparison of Resource Supply, Performance, Emissions, Fuel Storage, Safety, Costs, and Transitions*. Warrendale, Pennsylvania: Society of Automotive Engineering, 1988.

Federal Highway Administration. *Highway Statistics 1988*. Washington, DC: Government Printing Office # FHWA-PL-89-003, 1989.

Federation of American Scientists. *Preparing for the 1990s: The World Automotive Industry and Prospects for Future Fuel Economy Innovation in Light Vehicles*. Washington, DC: January 1987.

Gas Research Institute. *A White Paper: Preliminary Assessment of LNG Vehicle Technology, Economics, and Safety Issues.* Chicago, Illinois: January 10, 1992.

Gas Research Institute. *1992 Edition of the GRI Baseline Projection of US Energy Supply and Demand to 2010.* Chicago, Illinois: April 1992.

Gordon, Deborah. *Steering a New Course: Transportation, Energy, and the Environment.* Cambridge, Massachusetts: Union of Concerned Scientists, 1991.

INFORM (James S. Cannon). *Drive for Clean Air: Natural Gas and Methanol Vehicles.* New York: 1989.

Interagency Commission on Alternative Motor Fuels. *First Interim Report of the Interagency Commission on Alternative Motor Fuels.* Washington, DC: September 30, 1990.

Interagency Commission on Alternative Motor Fuels. *Second Interim Report of the Interagency Commission on Alternative Motor Fuels.* Washington, DC: September 1991.

International Association for Natural Gas Vehicles. *Natural Gas Vehicles: 1990.* Auckland, New Zealand: 1990.

International Association for Natural Gas Vehicles. *Conference Proceedings NGV90: 2nd International Conference and Exhibition, Buenos Aires, Argentina, October 21-25, 1990.* Auckland, New Zealand: 1990.

International Association for Natural Gas Vehicles. *Conference Proceedings NGV92: 3rd Biennial International Conference and Exhibition, Goteborg, Sweden, September 22-25, 1992.* Auckland, New Zealand: 1992.

International Energy Agency. *Energy Policies and Programmes: 1989 Review*. Paris, France: 1990.

International Energy Agency. *Substitute Fuels for Road Transport*. Paris, France: 1990.

International Energy Agency. *Greenhouse Gas Emissions: The Energy Dimension*. Paris, France: 1991.

International Energy Agency. *Energy Prices and Taxes: Fourth Quarter 1991*. Paris, France: 1992.

International Energy Agency. *Methane as a Motor Fuel*. Paris, France: May 1992. (Prepared by Sypher:Mueller International, Inc.)

MacKenzie, James. *Breathing Easier: Taking Action on Climate Change, Air Pollution, and Energy Insecurity*. Washington, DC: World Resources Institute, December 1988.

MacKenzie, James, and Michael Walsh. *Driving Forces: Motor Vehicle Trends and Their Implications for Global Warming, Energy Strategies, and Transportation Planning*. Washington, DC: World Resources Institute, December 1990.

Moreno, Rene and D. G. Bailey. *Alternative Transport Fuels from Natural Gas. World Bank Technical Paper #98*. Washington, DC: World Bank Publications, June 1989.

Motor Vehicle Manufacturers Association. *Facts & Figures '91*. Detroit, Michigan: 1992.

Natural Gas Vehicle Coalition. *Strategic Plan: June 1, 1990*. Washington, DC: 1990.

Oak Ridge National Laboratory. *Transportation Energy Data Book: Edition 10*. Oak Ridge, Tennessee: ORNL-6565, September 1989.

Office of Technology Assessment. *Replacing Gasoline: Alternative Fuels for Light-Duty Vehicles.* Washington, DC: US Congress, 1990.

Office of Technology Assessment. *Changing by Degrees: Steps to Reduce Greenhouse Gases.* Washington, DC: US Congress, 1991.

Ogden, Joan, and Robert Williams. *Solar Hydrogen: Moving Beyond Fossil Fuels.* Washington, DC: World Resources Institute, 1989.

Oppenheimer, Ernest. *Natural Gas: The Best Energy Source.* New York: Pen & Podium, Inc., 1989.

Renner, Michael. *Rethinking the Role of the Automobile.* Washington, DC: Worldwatch Institute, June 1988.

Samsa, Michael. *Potential for Compressed Natural Gas Vehicles in Centrally-Fueled Automobile, Truck and Bus Fleet Applications.* Chicago, Illinois: Gas Research Institute, June 1991.

Santini, D. J., et al. *Greenhouse Gas Emissions from Selected Alternative Transportation Fuel Market Niches.* Argonne, Illinois: US Department of Energy Center for Transportation Research, Argonne National Laboratory, 1989.

Sperling, Daniel, ed. *Alternative Transportation Fuels.* New York: Quorum Books, 1989.

Sperling, Daniel. *New Transportation Fuels: A Strategic Approach to Technological Change.* Berkeley, California: University of California Press, 1988.

US Department of Energy. *Assessment of Costs and Benefits of Flexible and Alternative Fuel Use in the US Transportation Sector.* Washington, DC: Four reports published 1988-1990.

US Department of Energy. *An Assessment of the Natural Gas Resource Base of the United States.* Washington, DC: May 1988.

US Department of Energy. *Alternative Fuel Vehicles for the Federal Fleet: Results of the 5-Year Planning Process.* Washington, DC: August 1992.

US Department of Transportation. *Moving America: New Directions, New Opportunities.* Washington, DC: February 1990.

US Energy Information Administration. *Annual Energy Review 1991.* Washington, DC: 1992.

US Energy Information Administration. *International Energy Annual 1991.* Washington, DC: 1992.

US Environmental Protection Agency. *Air Quality Benefits of Alternative Fuels.* Ann Arbor, Michigan: Prepared for the Alternative Fuels Working Group of the President's Task Force on Regulatory Relief, July 1987.

US Environmental Protection Agency. *Guidance on Estimating Motor Vehicle Emission Reductions from the Use of Alternative Fuels and Fuel Blends.* Ann Arbor, Michigan: January 1988.

US Environmental Protection Agency. *Analysis of the Economic and Environmental Effects of Compressed Natural Gas as a Vehicle Fuel.* Ann Arbor, Michigan: April 1990.

US General Accounting Office. *Alternative Fuels: Experiences of Brazil, Canada and New Zealand in Using Alternative Motor Fuels.* Washington, DC: GAO/RCED-92-119, May 1992.

Western Interstate Energy Board. *Alternative Fuel Use in Motor Vehicles: An Air Quality Perspective.* Denver, Colorado: April 15, 1987.

Glossary

Aftermarket conversion Retrofitting conventional factory-built vehicles with equipment for storing and burning an alternative fuel.

Attainment zone Region where air quality meets National Ambient Air Quality Standards set by the US Environmental Protection Agency. (See non-attainment zone.)

Bi-fuel engines Engines that can burn either an alternative fuel (e.g., natural gas) or a conventional fuel (e.g., gasoline or diesel), but not both simultaneously. (See dual-fuel engines.)

Big Three The largest United States automotive manufacturers (General Motors, Ford, and Chrysler.)

Commercialization With regard to natural gas vehicles, the selling for profit of fully warrantied vehicles capable of running satisfactorily on natural gas when used in normal driving patterns and vehicle applications.

Conversion kits Natural gas storage tanks, pressure regulators, fuel mixers, electronic control systems, tubes and fittings, and miscellaneous equipment needed to retrofit conventional vehicles to burn natural gas.

Dedicated engines Engines that burn only one fuel, e.g., natural gas.

Dual-fuel engines Engines that can burn two fuels simultaneously. (See bi-fuel engines.)

Early entry vehicle A vehicle produced in small numbers, yet commercially available.

Energy efficiency The effectiveness with which a source of energy is used to perform work; for example, a vehicle that burns fuel more efficiently can go farther using the same amount of energy than one that is less energy efficient.

Energy equivalency With regard to natural gas vehicles, the amount of natural gas that contains the same amount of energy as some amount of gasoline or diesel fuel.

Equivalent gallon The amount of natural gas containing the same energy as one gallon of a liquid fuel. Generally, about 115 standard cubic feet of natural gas contains energy equivalent to a gallon of gasoline.

Heavy-duty vehicles Vehicles (such as buses, refuse collectors, and semi-trailer trucks) generally weighing more than 14,000 pounds. The exact lower weight limit varies according to the definer of the vehicle category.

Hook-up The link that delivers natural gas from a pipeline to a home or business.

Incentive rates Rates for the sale of natural gas as a transportation fuel that are lower than the actual cost of producing and delivering the fuel to the end user; the additional costs are spread to other purchasers of natural gas in the area.

Lean burn Burning fuel with excess amounts of air to increase combustion efficiency, reduce the amount of fuel used, and lessen air pollution.

Light-duty vehicles Vehicles (including most cars and pickup trucks) that weigh less than about 8500 pounds. The exact upper weight limit varies according to the definer of the vehicle category.

Low-emission vehicles (LEVs) One of four categories of vehicles described in the California Air Resources Board's 1990 standards: LEVs must not emit more than 0.075 grams per mile of hydrocarbons.

Medium-duty vehicles Vehicles (including local delivery trucks, multi-passenger vans, and specialized service vehicles such as telephone repair trucks) generally weighing between 8500 and 14,000 pounds. The exact weight boundaries vary according to the definer of the vehicle category.

Non-attainment zone Region where air quality exceeds the National Ambient Air Quality Standards set by the US Environmental Protection Agency. (See attainment zone.)

Original equipment manufacturers (OEMs) With regard to the automotive industry, the original manufacturers or assemblers of factory-built vehicles, engines, or chassis. OEMs (for example, the Big Three auto producers) are differentiated from the manufacturers of individual automotive parts or the manufacturers of aftermarket conversion kits.

Otto cycle engines Conventional spark-ignited engines (as opposed to compression-ignited engines, such as diesel-powered engines).

Rate base The sum of "reasonable" capital investments on which utility companies are allowed, by public utility commissions, to calculate their rates.

Transitional low-emission vehicles (TLEVs) One of four categories of vehicles described in the California Air Resources Board's 1990 standards: TLEVs must not emit more than 0.125 grams per mile of hydrocarbons.

Ultra-low-emission vehicle (ULEVs) One of four categories of vehicles described in the California Air Resources Board's 1990 standards: ULEVs must not emit more than 0.04 grams per mile of hydrocarbons.

Zero-emission vehicles (ZEVs) One of four categories of vehicles described in the California Air Resources Board's 1990 standards: ZEVs must not emit any hydrocarbons at all.

Index

F (continued)

G

O

P

About the Author: James S. Cannon

James Cannon, Senior Fellow at INFORM, is an internationally recognized researcher, author, and analyst on energy development, environmental protection, and related public policy issues. Long associated with INFORM, he is the author of *Drive for Clean Air* (1989), the influential study of natural gas and methanol as alternative vehicle fuels. He is currently researching the viability of natural gas as a bridge to the solar hydrogen economy.

In addition to this work for INFORM, Mr. Cannon serves as an Energy Policy Analyst for the New Mexico Energy, Minerals, and Natural Resources Department where his assignments include Director of the State Energy Policy Project. Previously, he has held employee and consultant positions with several not-for-profit organizations, federal and state agencies — including a seven-year association with the US Office of Technology Assessment — and private companies. Mr. Cannon is frequently called upon to testify on energy and environmental issues before governmental, academic, and public audiences.

The author of two INFORM studies on acid rain, *Controlling Acid Rain: A New View of Responsibility* (1987) and *Acid Rain and Energy: A Challenge for New Jersey* (1984), Mr. Cannon also wrote two INFORM studies of coal conversion options in New York and New Jersey. He was the coauthor of INFORM's *A Clear View: Guide to Industrial Pollution Control* (1975) and its 1976 report on alternative energy sources, *Energy Futures: Industry and the New Technologies.*

James Cannon received an A.B. in chemistry from Princeton University and an M.S. in biochemistry from the University of Pennsylvania.

INFORM Publications and Membership Information

Energy and Air Quality

Selected Publications

Drive For Clean Air: Natural Gas and Methanol Vehicles (James S. Cannon), 1989, 252 pp., $65.00.

Municipal Solid Waste

Selected Publications

Making Less Garbage: A Planning Guide for Communities (Bette K. Fishbein and Caroline Gelb), 1992, 180 pp., $30.00.

Business Recycling Manual (co-published with Recourse Systems, Inc.), 1991, 202 pp., $85.00.

Burning Garbage in the US: Practice vs. State of the Art (Marjorie J. Clarke, Maarten de Kadt, Ph.D., and David Saphire), 1991, 275 pp., $47.00.

Garbage Management in Japan: Leading the Way (Allen Hershkowitz, Ph.D., and Eugene Salerni, Ph.D.), 1987, 152 pp., $15.00.

Forthcoming Publications on Municipal Solid Waste

Cleaner Products Study 1: Refillable Beverage Containers (working title)

Cleaner Products Study 2: Reusable Shipping Containers (working title)

Chemical Hazards Prevention

Selected Publications

Preventing Industrial Toxic Hazards: A Guide for Communities (Marian Wise and Lauren Kenworthy), available late spring/early summer 1993, ca. 140 pp., $25.00.

Environmental Dividends: Cutting More Chemical Wastes (Mark H. Dorfman, Warren R. Muir, Ph.D., and Catherine G. Miller, Ph.D.), 1992, 288 pp., $75.00.

Tackling Toxics in Everyday Products: A Directory of Organizations (Nancy Lilienthal, Michèle Ascione, Adam Flint), 1992, 180 pp., $19.95.

Cutting Chemical Wastes: What 29 Organic Chemical Plants Are Doing to Reduce Hazardous Wastes (David J. Sarokin, Warren Muir, Ph.D., Catherine G. Miller, Ph.D., and Sebastian R. Sperber), 1985, 548 pp., $47.00.

Other INFORM Publications

For a complete publications list, including materials on land and water conservation, and a quarterly newsletter, or for more information, call or write to INFORM.

Sales Information

Payment
Payment, including shipping and handling charges, must be in US funds drawn on a US bank and must accompany all orders. Please make checks payable to INFORM and mail to:

INFORM, Inc.
381 Park Avenue South
New York, NY 10016-8806
Tel (212) 689-4040

Please include a street address; UPS cannot deliver to a box number.

Shipping Fees
To order in the US, please send a check that includes $3.00 for the first book and $1.00 for each additional book for shipping and handling charges. To order in Canada, add $5.00 for the first book and $3.00 for each additional book. For information on shipping rates for other countries, call (212) 689-4040.

Discount Policy
Booksellers: 20% off on 1-4 copies of same title; 30% off on 5 or more copies of same title.

General bulk: 20% off on 5 or more copies of same title.

Public interest and community groups: books under $10 are not discounted; books $10-$25 cost $10; books $25 and up cost $15.

Returns
Booksellers may return books, if in saleable condition, for full credit or cash refund up to 6 months from date of invoice. Books must be returned prepaid and include a copy of the invoice or packing list showing invoice number, date, list price, and original discount.

Membership
Individuals provide an important source of support to INFORM and receive the following benefits:

Member ($25): A one-year subscription to *INFORM Reports*, INFORM's quarterly newsletter, and early notice of new publications.

Friend ($50): Member's benefits and a string bag.

Contributor ($100): Friend's benefits, plus a 10% discount on new INFORM studies.

Supporter ($250): Friend's benefits, plus a 20% discount on new INFORM studies.

Donor ($500): Friend's benefits, plus a 30% discount on new INFORM studies.

Associate ($1000): Friend's benefits, plus a complimentary copy of new INFORM studies.

Benefactor ($5000): Friend's benefits, plus a complimentary copy of new INFORM studies.